2013
万达商业规划

持有类物业 下册 VOL.2

WANDA COMMERCIAL PLANNING 2013
PROPERTIES FOR HOLDING

万达商业规划研究院 主编

中国建筑工业出版社

开福金街

万达广场
WANDA PLAZA
万达百货
1

EDITORIAL BOARD MEMBERS
编委会成员

萬達商業規劃

CONTENTS
目录

PART A FOREWORD
序言 370

PART B WANDA PLAZAS
万达广场 378

PART **C**
RESORT HOTELS 度假酒店

PART **D**
DESIGN & CONTROL 设计及管控

PART **E**
INDEX OF PROJECT 项目索引

PART

A

WANDA COMMERCIAL PLANNING

FOREWORD

序言

IMPROVE HOTEL DESIGN QUALITY AND BUILD INTERNATIONAL BRAND

推进酒店设计质量 打造酒店国际品牌

2013年万达集团逆势增长，在中国经济增速持续放缓的环境下，取得非常好的成绩。万达规模越来越大，业务范围越来越广，业态越来越丰富。2013年，中央出台“八项规定”，加上经济发展放缓，各种因素综合导致全国五星级酒店餐饮收入平均下降35%。在这种形势下，酒店管理公司今年共收入36.3亿元，同比增长30%，实属难能可贵，这是奋斗的结果。

2013年，酒店建设公司新开业16家酒店，其中五星和超五星级酒店13家，新增持有物业66.6万平方米，新增客房5339间，总客房数达17017间；这一年，酒店管理公司多次召开设计标准研讨会，就酒店设计标准的架构、内容及执行方式深入地进行互动交流及经验分享，并形成设计规划改进销项计划，进一步推进了万达酒店规划设计的质量。

同时万达集团在文化旅游产业上也有了突破性的发展。2013年开工哈尔滨、南昌和合肥万达城。三个万达城项目开工，震动中国，拉开“万达茂”在中国全面布局的大幕。这几个项目各具特色，都有很多创新。

2013年万达收购圣汐游艇公司和投资伦敦豪华酒店，在跨国发展上也震惊了世界。万达投资7亿英镑在泰晤士河边建设酒店和公寓综合项目。项目位置绝佳，位于伦敦第二富人区，可以俯瞰伦敦海德公园和白金汉宫。公寓高度200米，是伦敦最高的住宅。酒店有188个房间，建成后绝对是欧洲第一奢华酒店。

2014年是万达标志年，当前的社会环境、资源、舆论等各种条件都有利于万达，可以说万达正处于历史发展的最好时期。万达酒店计划新开业18个五星和超五星酒店，新增客房5485间，累计开业酒店72家。酒店管理公司收入47.8亿元，依靠科技和创新继续提升酒店设计、建设和管理的品质。

我有一个梦想，不光把企业做大，还要把中国酒店品牌打到全世界。我也知道，做豪华酒店这种超级奢侈品，至少也要二三十年的时间建立这个品牌，所以很多企业一看要二三十年就不做了，那我来做这先行者，我来熬这个前面的二十年，一定要把中国的酒店品牌打到世界上去。

——摘编自王健林
《万达集团2013年工作总结暨2014年工作安排》

In 2013, under a continuous decelerated growth of Chinese economy, Wanda Group performed well with excellent achievements. The scale grew, the business scope expanded and the commercial activities became more and more rich. In the same year, the Political Bureau of the Communist Party of China (CPC) Central Committee issued the "Eight-point Regulation" and the economic development slowed down, various factors led to a 35% decrease of 5-star hotels catering earning on average all over China. However, Wanda Hotels & Resorts earned RMB 3.63 billion in total, with year-on-year growth of 30%. It's really a rare and praiseworthy achievement, and an outcome of persistent efforts!

Wanda Hotel newly opened 16 hotels in 2013, including 13 five-star and super five-star hotels, and increases new properties for holdings of 666,000 m^2 and new guestrooms of 5,339 ones, reaching a total number of 17,017 guestrooms. In this year, Wanda Hotels & Resorts held several workshops on design standard, during which in-depth interaction and experience sharing of hotel design standard structure, content and execution mode were actively discussed and design & planning improvement plan was established, further improved the planning and design quality of Wanda hotels.

Meanwhile, Wanda Group had also made breakthroughs on the cultural tourism industry. In 2013, Wanda has begun construction on Cultural Tourism City developments in Harbin, Nanchang and Hefei. Simultaneous commencement of these three projects shakes the whole country and officially starts the blossoming of Wanda Mall all over China. These three distinctive cultural tourism cities represent various innovations of Wanda.

Moreover, Wanda Group also made achievement on multinational development, the acquisition of Sunseeker and investment on luxury hotels in London in 2013 shocked the world. Wanda invested 700 million pounds in constructing a hotel and apartment complex by the Thames in London. The project enjoys excellent location, belonging to the second wealthy area, where guests can have a panoramic view of Hyde Park and Buckingham Palace. The complex is 200m high and will be the highest residential building in London. With 188 rooms, the hotel will undoubtedly be the top luxury hotel in Europe when it is completed.

The Year 2014 is a symbolic year for Wanda Group, the present social environment, resource, public opinion and other conditions are beneficial for Wanda, in a manner of speaking; Wanda Group is in its best stage of historical development. Wanda Hotel plans to newly open eighteen five-star and super five-star hotels, reaching a total number of 72 hotels and increase 5,485 guest rooms. Wanda Hotels & Resorts plans to reach an income of RMB 4.78 billion and further improve design, construction and management quality relying on science & technology and innovation.

I have a dream that besides expanding the business, I will make a world-known Chinese hotel brand. I know that making luxury hotels, a kind of super luxury industry, will take at least two or three decades to establish a brand, and that's why many enterprises have given up on this way. I mean to be the pioneer and endure the first twenty years, to make Chinese hotel brand onto the world's market!

Quoted from *Annual Work Report 2013 and Work Plan 2014 of Wanda Group*
by Chairman Wang Jianlin

万达集团董事长
王健林
Wang Jianlin
Chairman of Wanda Group

WANDA GREEN BUILDING STRATEGIC GOAL AND IMPLEMENTATION CONTROL

万达绿色建筑战略目标与实施管控

文 / 万达商业规划研究院常务副院长 叶宇峰

2013年是万达集团成立25周年，也是万达商业规划院研究院注册成立6周年，截至2013年12月31日，累计114个项目（含51个万达广场、26个酒店、35个住宅项目、1个万达学院、1个文旅项目）取得“绿色建筑设计标识”认证，累计29个项目（含28个万达广场、1个万达学院）取得“绿色建筑运行标识”认证，19个酒店取得“绿色饭店运行标识”认证。万达集团已经成为国内获得绿色建筑标识最多的企业（图1）。

（图1）绿色建筑标识证书

一、绿建战略目标

万达集团的节能工作战略目标为：

1 商业建筑——引领行业发展

2011年及以后开业的项目均取得绿色建筑1星设计标识
2011年至2015年间新开业项目逐年降低运行能耗2%~3%
2013年2个项目取得绿色建筑1星运营标识认证
2015年实现运营管理水平均达到绿色建筑1星运营标准

2 酒店建筑——行业领先

2011年至2015年间新开业项目逐年降低运行能耗2%~3%
2015年以前取得5个绿色建筑1星设计标识
2015年实现运营管理水平均达到绿色饭店金叶级运营标准

3 居住建筑——行业领先

2012年及以后所有居住建筑均取得绿色建筑1星设计标识
2013年及以后的住宅产品均为精装修交付

在2011年召开的第七届绿色建筑大会上，我们正式向社会宣布了万达集团绿建节能的战略目标，主动承担社会责任，接受社会各界的监督。在《万达集团“绿色、低碳”战略研究报告》和《万达集团节能工作规划纲要》（2011-2015）的指导下，2010年开业的广州白云万达广场成为我国首个获得“绿建设计

The Year 2013 is the 25th anniversary of Wanda Group and also the 6th anniversary of Wanda Commercial Planning & Research Institute. As of December 31st, 2013, 114 projects in total, including 51 Wanda Plazas, 26 Hotels, 35 residential projects, one Wanda Institute and one Culture & Tourism project, have obtained the certification of Green Building Design Label; 29 projects, including 28 Wanda Plazas and one Wanda Institute, have obtained the certification of Green Building Operation Label and 19 hotels have obtained the Green Hotel Operating Label Certification. Wanda Group has become the enterprise that has obtained the largest number of green building labels in China (Fig. 1).

I. GREEN BUILDING STRATEGIC GOAL

The strategic goal of energy conservation of Wanda Group is:

1. COMMERCIAL BUILDINGS - LEADING DEVELOPMENT OF THE INDUSTRY

The projects opened in and after 2011 shall all obtain the green building one star design label
The newly opened projects from 2011 to 2015 shall reduce operating energy consumption by 2%~3% year by year
Two projects shall obtain the green building one star operation label certification in 2013
The operation and management shall all reach green building one star operating standard in 2015

2. HOTEL BUILDINGS - THE INDUSTRY LEADER

The newly opened projects from 2011 to 2015 shall reduce operating energy consumption by 2%~3% year by year
Five green building one star design labels shall be obtained before 2015
The operation and management shall all reach green hotel gold leaf level operating standard in 2015

3. RESIDENTIAL BUILDINGS - THE INDUSTRY LEADER

The projects opened in and after 2012 shall all obtain the green building one star design label
All residential projects in and after 2013 shall be delivered with fine decoration

On the 7th International Conference on Green and Energy-Efficient Building & New Technologies and Products Expo in 2011, Wanda Group has formally announced the strategic goal of green building and energy conservation of Wanda Group to the public

内部资料　档案汇编

万达集团“绿色、低碳”战略研究报告

WF-2010-12

2010.10

大连万达商业地产股份有限公司
DALIAN WANDA COMMERCIAL ESTATE CO., LTD.

内部资料　档案汇编

万达集团节能工作规划纲要

（2011-2015 年）

2011.12

大连万达商业地产股份有限公司
DALIAN WANDA COMMERCIAL ESTATE CO., LTD.

（图2）战略研究报告与规划纲要

标识”2星的大型商业项目，2011年开业的13个万达广场及8个住宅项目取得了“绿建设计标识”，万达学院成为国内首个获得3星级“绿建设计标识”的校园园区项目，到2012年当年开业商业项目全部获得设计认证，并有10个商业项目获得“绿建运行标识”认证，实现了国内大型商业购物中心“绿色建筑运行标识”零的突破，真正实现了大型商业建筑从设计到运行的全过程绿建节能。2012年能源管理平台全部正式上线运行，至2013年年底已覆盖北京石景山、上海江桥、武汉菱角湖等38个万达广场，并在集团设立了总部管理平台系统。节能效果显著，各地万达广场平均能耗2012年比2011年降低10%，2013年比2012年降低5.2%，远超原定的2%~3%的节能战略目标。（图2）

二、实施管控措施

万达集团能够按照预定的战略目标兑现承诺靠的是组织保障、制度落实、计划管理和信息化监管等4个方面的管控措施。

1 组织保障

万达集团成立了节能工作管理小组，是集团节能工作的领导机构。成立于2010年的万达商业规划研究院节能所，是集团绿建节能工作技术归口管理部门，由一批绿建节能专家和技术人员组成，专项负责集团绿建节能方面的科研、实施和管理，重点对集团节能战略目标的制定和实施进行计划管理及督办（图3）。

and expressed our willing to shoulder the social responsibilities and accept oversight of the public. Guided by *Wanda Group's "Green, Low-carbon" Strategy Research Report* and *Wanda Group Energy Conservation Planning Work Program (2011-2015)*, Guangzhou Baiyun Wanda Plaza, which opened in 2010, becomes the first large-scale commercial project which obtained the two-star Green Building Design Label in China, the thirteen Wanda Plazas and eight residential projects, opened in 2011, obtain the certification of Green Building Design Label, and the Wanda Institute becomes the first campus project that obtained the three-star Green Building Design Label in China. In 2012, newly opened commercial projects all obtain the design certification, and ten of which even obtain the certification of Green Building Operation Label, symbolizing a zero breakthrough of Green Building Operation Label for large-scale commercial shopping center in China and realizing green building and energy conservation of large-scale commercial buildings during whole process from design to operation. In the same year, the energy management platform has been formally put into operation and rapidly covered Beijing Shijingshan, Shanghai Jiangqiao, Wuhan Lingjiaohu and other thirty five Wanda Plazas at the end of 2013. The headquarter management platform system is also set in the Group, and significant energy conservation effects have been achieved. Compared with energy consumption of the previous year, the average energy consumption of Wanda Plazas all over China reduces 10% in 2012 and 5.2% in 2013, far more than the original strategic goal of 2%~3% (Fig. 2).

II. IMPLEMENTATION CONTROL MEASURES

It is with organizational guarantee, system implementation, plan management and information supervision, these four management and control measures, can Wanda Group honor its commitment to fully complete the intended strategic goal.

1. ORGANIZATIONAL GUARANTEE

Wanda Group established an energy saving management team which is a directing organization for energy conservation of Wanda Group. The Energy Saving Research Institute of Wanda Commercial Planning Institute, established in 2010, is a technical management team for energy saving work and composed of a group of experts and technicians in this field. The management team is especially

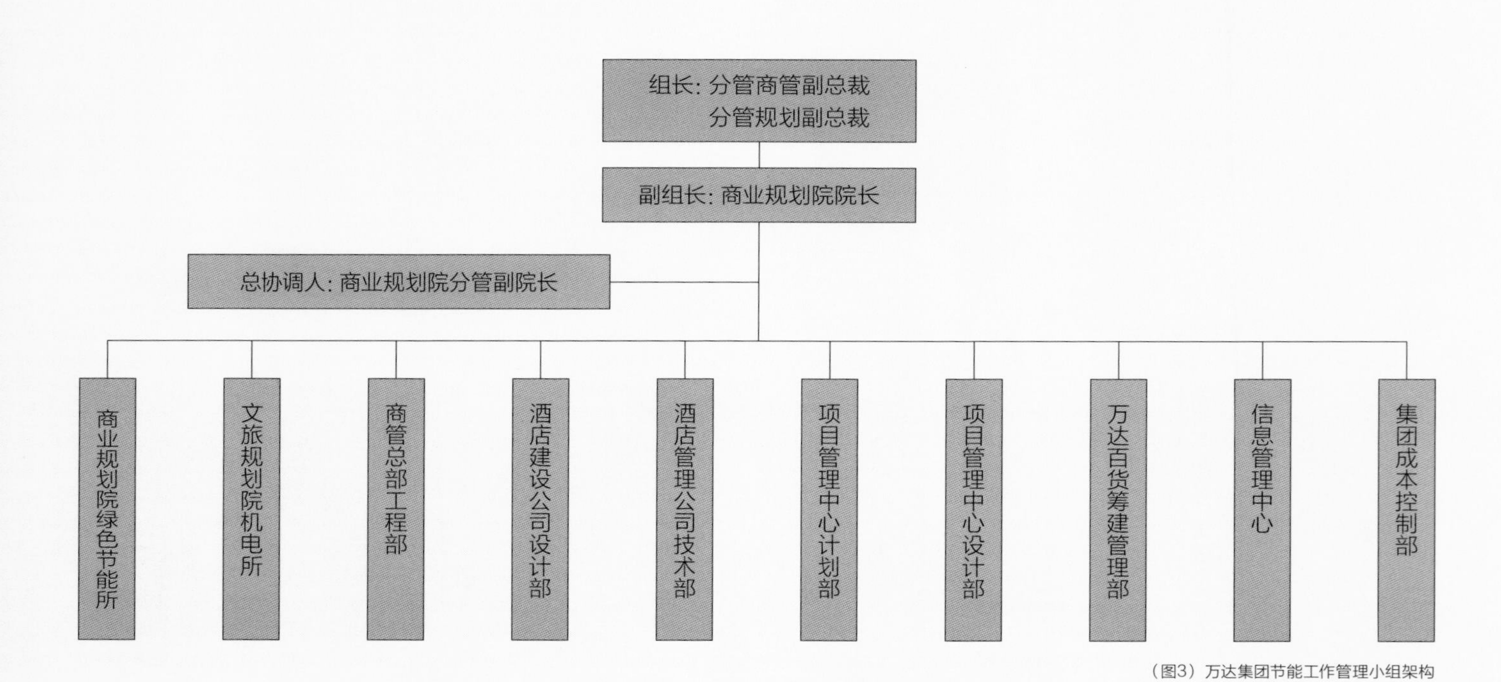

（图3）万达集团节能工作管理小组架构

2 制度落实

绿建节能工作已经全面纳入集团2012版《项目设计管理制度》，各部门有关绿建节能的工作职责、业务事项、成果形式、工作流程等都在制度中明确（图4）。

序号	业务实现	操作流程		流程附件
		发起流程	审批、接收部门	
7.2	绿建设计星级评定成果备案	项目公司设计部—规划副总—总经理	中心计划部项目负责人、规划院项目负责人、节能所所长、建筑分管副院长、商管公司、商管总部、酒店总部（有酒店项目）	设计标识证书及申报资料

（表4）《项目设计管理制度》商业建筑的节能工作设计管理流程

3 计划管理

在万达集团项目管理的计划模块化中2个2级节点，1个3级节点与绿建节能有关，所有项目的绿建节能工作全部纳入计划模块化管理（图5）。

4 信息化监管

在万达《综合体项目计划模块化管理办法》中，已经纳入绿建节能相关的工作节点。每个万达广场的建设现场均安装了可视化信息采集装置，可以对节能工程的实施情况进行在线监控。

responsible for scientific research, implementation and management of green building energy saving work, and focusing on plan management and supervision on making and implementing energy saving strategic goals of Wanda Group (Figure 3).

2. SYSTEM IMPLEMENTATION

The energy saving work for green buildings has been fully incorporated in *Project Design Management System 2012*, in which duties and responsibilities, assignments, outcomes and work flows of energy saving work are expressly specified (Fig. 4).

3. PLAN MANAGEMENT

Among the project management modules of Wanda Group, 2 two-tier joints and 1 three-tier joints are concerned with energy conservation for green buildings, and energy saving work of all projects is all incorporated in the plan module management (Fig. 5).

序号	级别	阶段	事项	开始时间	周期	完成时间
296	2	验收	绿色建筑设计评星成果取得	开业日前210天	150	开业日前60天
315	2	验收	节能相关工程实施成果专项验收	开业日后150天	30	开业日后180天

（表5）有关绿建节能工作的计划模块节点

万达集团作为中国民营企业的代表，以争做百年企业为理想，做事目光长远，追求长期利益，积极参与国家倡导的节能环保事业。从2011年起，已连续3年参加“国际绿色建筑与建筑节能大会”，并2次主办“大型公共建筑的节能运行与监管”分论坛，发行4本节能专刊，出版《绿色建筑——商业地产中绿色节能的实践及探索(一)》一书。今后，万达集团将会继续践行绿色建筑的承诺，向世界展示中国民营企业的社会责任感。

4. INFORMATION SUPERVISION

Work related to energy saving has been incorporated in Wanda's Plan Module Management Methods for Complex Projects. Construction site of each Wanda Plaza has been equipped with visual information collection device which can realize on-line supervision on implementation of energy saving works.

As representative of Chinese private enterprises, Wanda Group pursues to be a 100-year enterprise with its focus on long term vision and benefit and has been actively engaged in the energy saving and environmental protection work which is advocated by China. Since 2011, Wanda Group has been successively for three years participated in "International Conference on Green and Energy-efficient Building", hosted 2 times of the forums of "Energy Saving Operation and Supervision for Large Public Buildings", published 4 special issues concerning energy saving work, and published the book *Green Building Vol. 1 - Exploration of Energy Conservation and Green Practice in Commercial Buildings*. In the future, Wanda Group will continue to practice its green building commitment and show the world its social responsibility of being a Chinese private enterprise.

审图号：GS（2014）1915号

PART

B WANDA COMMERCIAL PLANNING WANDA PLAZAS 万达广场

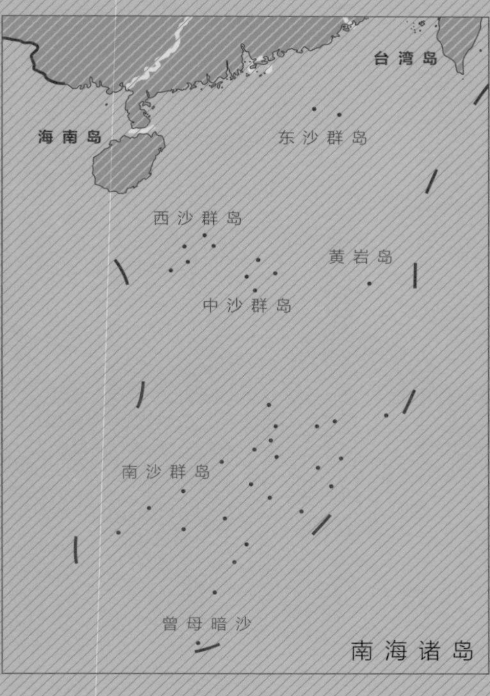

CHANGSHA KAIFU WANDA PLAZA
长沙开福万达广场

开业时间 2013 / 09 / 27
建设地点 湖南 / 长沙
占地面积 12.18 公顷
建筑面积 103.69 万平方米

OPENED ON SEPTEMBER 27 / 2013
LOCATION CHANGSHA / HUNAN PROVINCE
LAND AREA 12.18 HECTARES
FLOOR AREA 1,036,900 m^2

OVERVIEW OF PLAZA
广场概述

长沙开福万达广场地处长沙市繁华核心区，依水而建，俯瞰橘子洲，地理位置得天独厚。基地周围环绕城市主干道，周边环境成熟，商业氛围浓厚，交通十分便捷。长沙开福万达广场是万达集团在长沙市投资建设的首个大型城市综合体，也是集团首个开业的奢侈品店定位的广场。项目由超A级购物中心、超五星级酒店、甲级写字楼群、室外步行街和豪宅等组成，总建筑面积超过103万平方米，其中地上建筑面积约80万平方米，地下约23万平方米，业态丰富、规模宏大，是万达集团在长沙市的扛鼎之作！

Located in the central bustling area of Changsha city, Changsha Kaifu Wanda Plaza is constructed by the river to overlook the Orange Isle and enjoys an exceptional geographical location. Surrounded by urban main roads, the site is benefited by the ripe surroundings, strong commercial atmosphere and convenient transportation. Kaifu Wanda Plaza is the first large-scale urban complex invested and constructed by Wanda Group in Changsha, and also the first opened luxury stores-oriented plaza of Wanda Group. As a complex consists of super A-class shopping center, super five-star hotel, grade-A office complex, exterior pedestrian street and luxury housing, Changsha Kaifu Wanda Plaza covers a gross floor area of 1.03 million square meters, including around 800,000 m^2 for aboveground area and 230,000 m^2 for the basement. Concerning various commercial activities and a grand scale, Changsha Kaifu Wanda Plaza is a representative work of Wanda Group in Changsha, Hunan province.

1

1 广场总平面图
2 广场全景

2

万达广场
PLAZA

FACADE OF PLAZA
广场外装

现代、简约、刚劲、大气，是长沙开福万达广场建筑群的风格特征。黑白灰色调与简洁的线条经过反复推敲，摒弃冗余，刻画了高端的品质定位。建筑立面以简约竖向线条统一建筑群体形象，以雄健挺拔之势俯瞰湘江。立面构造与夜景照明完美结合，全天候展现设计之精彩。

The building complex in Changsha Kaifu Wanda Plaza is modern, simple, vigorous and grand, adopting black and white gray tone and concise lines and presenting high quality of the hotel. Building elevations are expressed in simple vertical lines, forming a robust and lofty building complex image overlooking the Xiangjiang River. Elevation structure is in perfect harmony with nightscape lighting, thus showcasing wonderful design feature all the time.

4

3 广场外立面
4 广场仰视

5 广场百货外立面

万达影城
WANDA CINEMA
IMAX
华润万家
城市超市
ESTEE LAUDER
雅诗兰黛
CLARINS
Double Serum

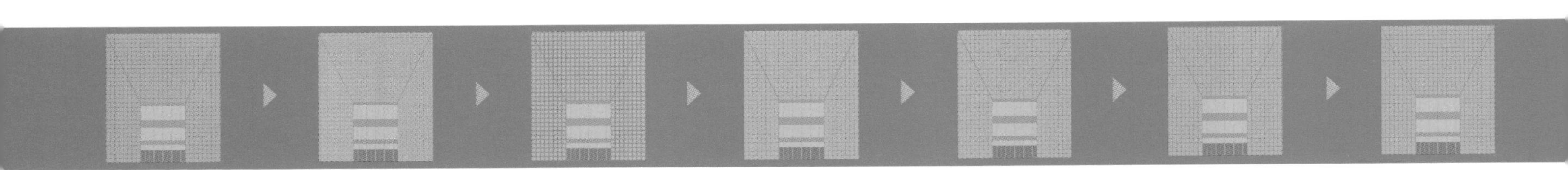

6

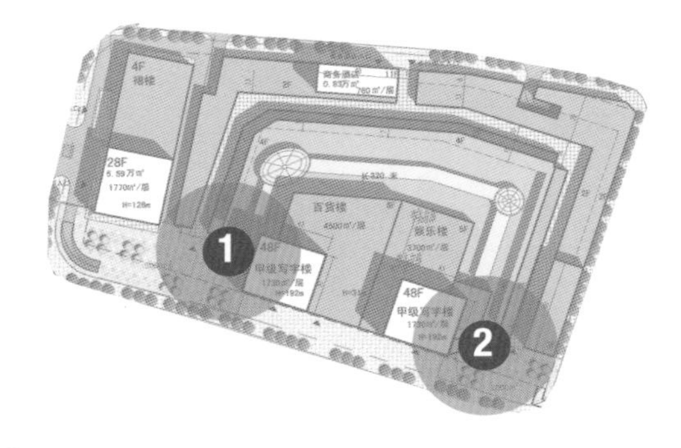

7

1

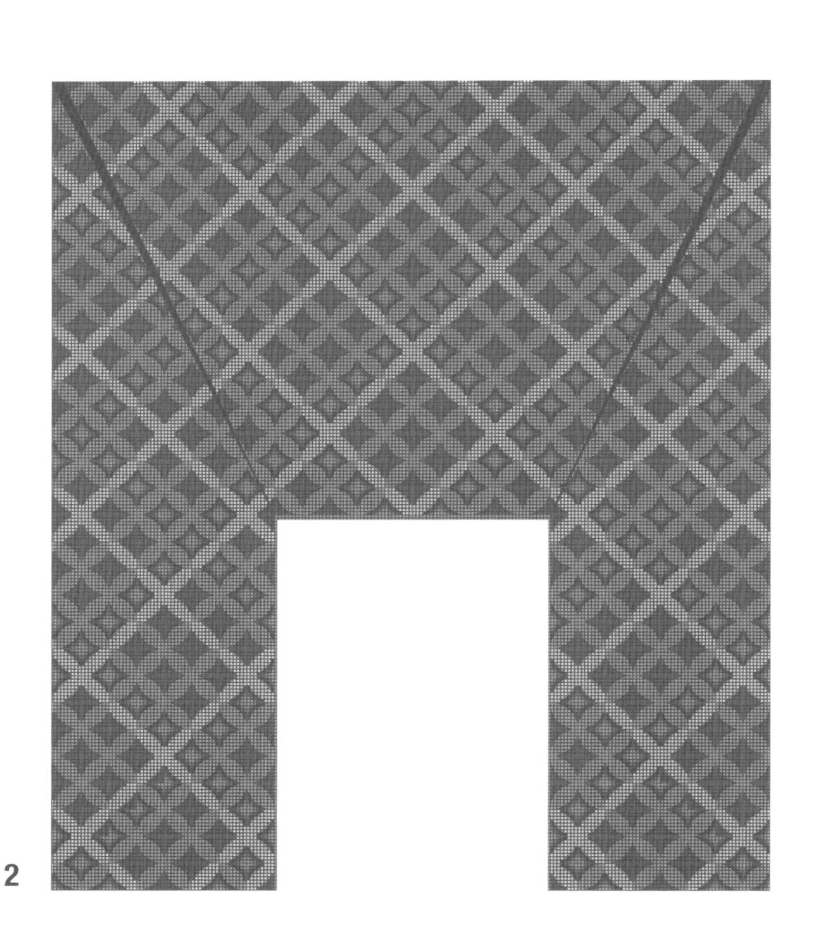

2

8

9

超大门头、特种玻璃设计、高清细腻夜景效果、奢侈品店定位门头肌理，呈现高端奢华入口商业形象，成为长沙开福万达广场画龙点睛之笔。

Super large gate, special glass, high-definition and delicate nightscape design effect and luxury store-oriented gate texture, all of these showcase commercial image of the high-end and luxurious entrance, and is the finishing touch for Changsha Kaifu Wanda Plaza design.

6 门头设计过程方案
7 一、二号门头最终方案
8 广场一号入口
9 广场二号入口

10 广场主入口

万达广场
WANDA PLAZA
万达百货
WANDA DEPT.STORE

collncos

11

12

INTERIOR OF PLAZA
广场内装

长沙开福万达广场设计首次实现万达集团内，万达百货与室外步行街整体效果管控，内场效果高档时尚，简约明快，恰如其分地营造了高端购物氛围。椭圆中庭的两边侧裙板采用了跟圆中庭一样的做法，而在观光电梯的设计上，沿用了“金缕玉衣”的外形设计并加以延伸，电梯的整体装饰材料还采用了玻璃幕墙，在外层包裹了一层金属装饰网，犹如给观光梯披上了一件金纱，既富有动感，更是用现代手法描绘了“湖湘文化”的深奥之处。

In the Plaza design, overall effect of Wanda department store in the exterior pedestrian street is under central control within Wanda Group. The interior finishing is upscale, fashionable, simple and concise, appropriately creating a high-end shopping atmosphere. Plates on both sides of the oval atrium adopt the same method as that for side plates on the circular atrium. Panoramic lift shape is designed based on the "jade clothes sewn with gold wire", finished with glass curtain wall where one layer of metallic decoration net is wrapped on the external layer, which looks as if the panoramic lift is dressed with a gold yarn, thus making it dynamic and presenting the profound Huxiang Culture (Hunan Province Culture) with modern means.

11 椭圆中庭
12 圆中庭

圆中庭的两侧侧裙板，将钢化艺术玻璃作为侧裙板，其玻璃上的艺术纹理就采用漓湘水流的纹理——波浪，意喻漓湘水流奔腾不息。中庭的观光电梯是两边侧裙板的汇合点，在观光电梯的底端设计流水幕墙，喻示支流波涛汇入湘江。

Tempered art glass plates are set on both sides of the circular atrium to serve as barriers, and the glass plates are designed with texture of the Lijiang River and the Xiangjiang River-wave, implying that Li River and the Xiangjiang River roll ahead ceaselessly. Panoramic lift at the atrium is a point where side plates meet. Flowing water curtain wall at the bottom of the panoramic lift implies that all tributaries converge to the Xiangjiang River.

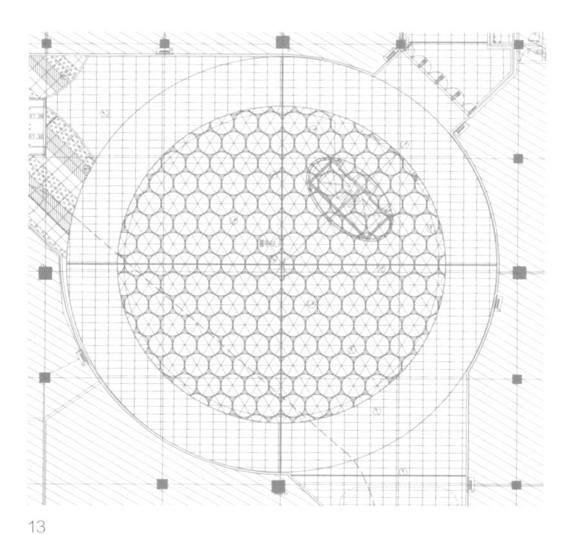
13

14

13 圆中庭采光顶平面图
14 圆中庭仰视
15 圆中庭

L&G
L&G
ZUMA
LUXURY
15

16 圆中庭观光电梯水景
17 室内步行街灯饰
16

广场圆中庭观光电梯首次引入水景设计，配合水幕喷头，营造出湘江水流入万达广场的特殊视觉效果并深受董事长好评，被誉为“匠心独韵”的设计。广场百货中庭首次由万达商业规划研究院管控设计，与步行街装修风格的一体化设计，注重细节装饰效果；首次在百货中庭设计巨型灯具营造优雅空间效果。

In design for panoramic lift at the circular atrium of the plaza, waterscape design introduced for the first time and water curtain nozzle jointly create a special visual effect that the Xiangjiang River water flows into the Plaza, which is praised by the president, and billed as an "original" design. Design and control of department store atrium by the planning institute for the first time is in harmony with the finishing style of the pedestrian street, focusing on detail finishing effect and creating an elegant spatial effect using giant lighting fixture in department store atrium design.

18

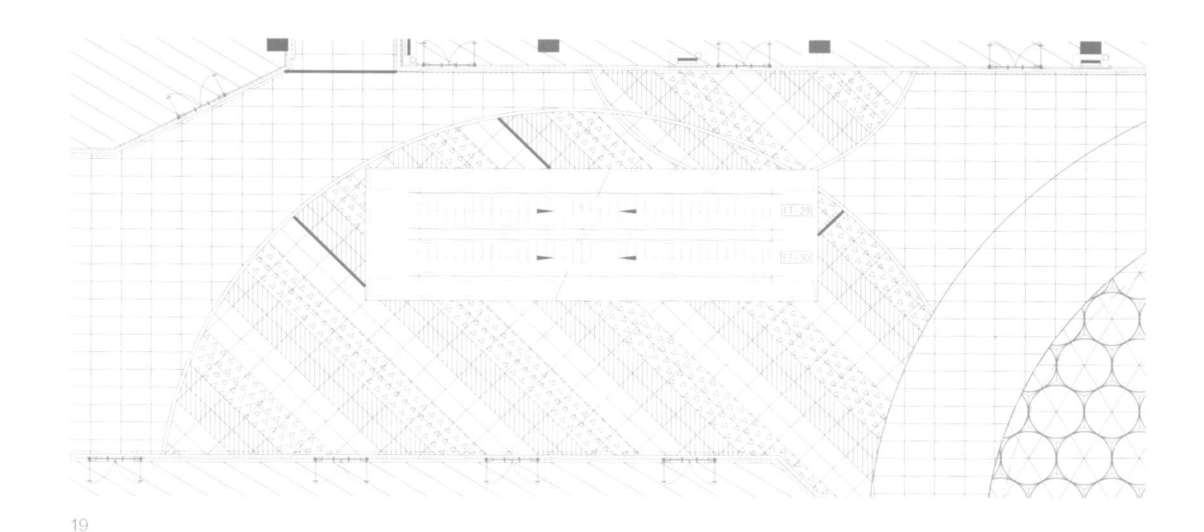

19

20

为体现湘江的蜿蜒及平畴万顷，在长街的设计上采用玻璃作为拦河侧裙板，顶层侧裙板更是采用建筑玻璃幕墙隐框式安装，背透光，搭配线性天花及建筑本身结构就如同湘江之水激流而下，奔腾不息。长街两边商铺的装饰更是为空间增添亮点，犹如两岸一派江南水乡景象，构成了美丽的长沙沿江风光带。

连桥的造型设计采用连桥底部中间突出，两侧向上倾斜的设计手法，寓意为不管湖南水系多广、支流多少，最终汇流于湘江；连桥底部突出部分就好似湘江，起到连接作用；而在材质的选择上采用微孔铝单板，打破了人们长期的固化思维，铝单板不仅可以用在室外，在室内空间也可展现别样风韵。

Glass side plates are utilized as barriers for long gallery to reflect winding but broad the Xiangjiang River; top layer side plates are concealed in glass curtain wall frames, with back light, and linear ceiling and the building structure create an image of torrent and surging the Xiangjiang River. Shops on both sides of the long gallery add colors to the space and form a beautiful scenic belt along rivers in Changsha.

Overhead bridge shape design is featured by protruded central part and upswept sides at the bridge bottom, implying that vast and various water systems in Hunan Province will finally converge to the Xiangjiang River. The protruded part at the overhead bridge bottom serves as a bridge like the Xiangjiang River. Adoption of porous aluminum veneers break habitual way of thinking, proving that aluminum veneers can not only be used outdoors, but also indoor to showcase a different charm.

18 室内步行街
19 室内步行街平面图
20 侧裙板细部

LANDSCAPE OF PLAZA
广场景观

景观环境设计理念秉承简约而现代的风格，建筑立面和景观环境全部以简洁的直线进行构图，通过不同的材料体块互相交织、穿插的设计手法，体现出高端城市综合体内在的复杂与有序并存的特性。

广场长方形种植池、坐凳和木平台的排列与地面铺装的排列方式一致，加强了建筑空间布局的完整性，并较好地解决了商业人流的休憩需要。整体方案以现代简约风格为基调，与立面相呼应，体现奢侈品店的定位。北广场主雕塑“精卫鸟”用解构主义手法表现开福广场的多姿与时尚，与湘江共歌，结合镜面水池成为主要标志。

The landscape design of plaza follows simple but modern principles. Building facades and landscape environment are formed by concise straight lines, through the method of inter-weaving different material blocks, thus harmonizing complexity and orderliness, inherent characteristics of high-end urban complex.

Rectangular planting bed, bench and wooden platform in the plaza are arranged in a way the same as that for floor pavement, making spatial layout more integrate and better meeting recreation demand of commercial pedestrian flow. The overall plan, modern and simple, corresponds to elevation design and reflects the position of luxurious store. The northern plaza is arranged with the main sculpture of "Bird Jingwei" made based on deconstruction, demonstrating a variable and fashionable plaza which is in harmony with the Xiangjiang River. The sculpture and mirror water pool become a main symbol.

23

21

22

21 广场喷泉
22 广场座椅
23 主雕塑

唯此可达

25

NIGHTSCAPE OF PLAZA
广场夜景

秉承“见光不见灯”的理念，通过独特的灯具发光面设计，既满足了夜间灯具表面发光亮度，又保证了白天观赏效果的完整性。以光为笔，浸蘸时尚奢华与地域人文，在湘江之畔描绘万达广场夜晚的韵味，成为一道城市之光，闪烁在星城。

超大高清显示门头设计是夜景照明的又一亮点，采用独特的装饰显示屏的设计手法，呈现白天与夜晚不同感受，“一静一动”相得益彰。在白天，入口门头大气而不张扬，超大体量的门头搭配细腻的玻璃肌理，使其独具奢华气质。

首次采用“裸眼 3D”动画制作技术，对建筑造型进行重新解构，通过制造奇特的三维变化，重新定义建筑空间感受。

Based on the concept of "lighting with invisible light", the nightscape design of the Plaza adopts unique light luminous surface design, which not only satisfies luminance requirement of lighting fixtures at night, but also guarantees integrate viewing effect at daytime. The Plaza lighting design showcases fashion, luxury and regional humanity and demonstrates lasting appeal of Wanda Plaza at night at the bank of the Xiangjiang River, which appears like twinkling light of the city.

Super-large and high definition gate design is another highlight of the nightscape lighting, adopting distinctive decoration display screen to bring about different feelings at daytime and nights, and bringing out the best in each other through "a combination of static and dynamic features". The gate is grand but not flaunt at daytime, and super-large gate and fine glass texture make it luxurious.

"Naked eye 3D" animation production technique is adopted for the first time to re-deconstruct the building shape, and re-define space perception of the building via unique 3D variations.

25 主入口液晶屏图案
26 主入口夜景

万达广场
WANDA PLAZA

27

27 外立面人像投影
28 外立面彩色格子投影
29 外立面夜景

WandaVista
WandaVista
万达广场
万达广场

OUTDOOR PEDESTRIAN STREET
室外步行街

“好看落日斜衔处，一片春岚映半环”。室外步行街以现代简约手法诠释商业街的特色。深色背景幕墙衬托现代彩色元素，如彩链绕颈，环绕购物中心，与之互映互衬，构成完整的商业视觉体验。

"Setting sun glowing at the mountain side, spring wind lighting up the other side of the sun". The outdoor pedestrian street design interprets characteristics of commercial street with a modern and simple method. Modern color elements are set against deep colored background curtain wall, embracing the shopping center like colorful chains surrounding neck, thus offering a complete commercial visual experience.

31 室外步行街外立面
32 景观雕塑
33 室外步行街入口

开福金街

FUSHUN WANDA PLAZA
抚顺万达广场

开业时间 2013 / 10 / 25
建设地点 辽宁 / 抚顺
占地面积 13.88 公顷
建筑面积 93.21 万平方米

OPENED ON OCTOBER 25 / 2013
LOCATION FUSHUN / LIAONING PROVINCE
LAND AREA 13.88 HECTARES
FLOOR AREA 932,100 m^2

OVERVIEW OF PLAZA
广场概述

抚顺万达广场项目位于辽宁抚顺市新抚区，位于浑河南路南、东一路轻轨以北、中央大街以东和东三街以西，浑河南侧，处于抚顺市中心位置。总规划用地面积 13.88 万平方米，总建筑面积 93.21 万平方米。地上建筑面积 75.2 万平方米（含购物中心 10.57 万平方米、酒店 3.23 万平方米、写字楼 13.64 万平方米、住宅 25.59 万平方米、商铺 6.15 万平方米，城市配套物业及回迁物业 16.02 万平方米）；地下建筑面积 18.01 万平方米。购物中心共五层，涵盖了万达百货、万达影城、大歌星 KTV、大玩家、乐天玛特超市、大型酒楼、健身及多个主力业态。

Fushun Wanda Plaza is located in Xinfu District of Fushun of Liaoning Province, to the south of South Hunhe Road, to the north of light rail along Dongyi Road, to the east of Zhongyang Avenue, to the west of Dongsan Street, and to the south side of Hun River. The plaza is situated in central Fushun, with a total planned land area of 138,800 m^2 and a gross floor area of 932,100 m^2. The floor area for aboveground part is 752,000 m^2 (including 105,700 square meters of shopping center, 32,300 square meters of hotel, 136,400 square meters of office building, 255,900 square meters of residence, 61,500 square meters of store area and 160,200 square meters of supporting urban property and move-back property) and the floor area for underground part is 180,100 m^2. The shopping center is a five-storey building containing such main business types as Wanda Department Store, Wanda Cinema, Big star KTV, Super Player Center, Lotte Matt Supermarket, large hotels, fitness and others.

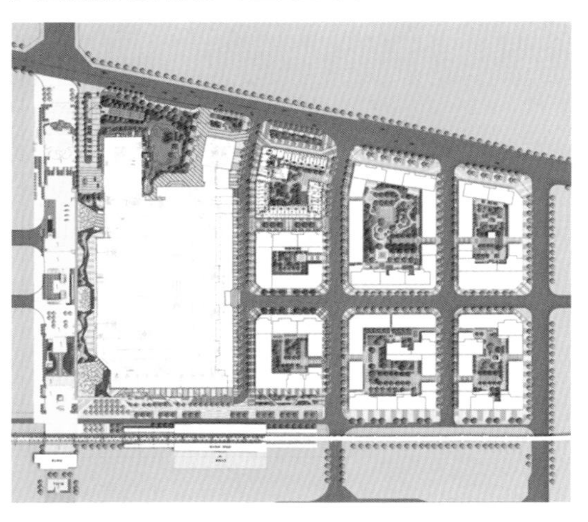

1

2

1 广场总平面图
2 广场鸟瞰图

3

FACADE OF PLAZA
广场外装

满清文化、矿产是抚顺这一城市的特点，设计从中汲取灵感，引入金属板材料，并精心设计了模块化的窗花纹理，达到古典与现代完美结合的效果。镂空金属板形成投影，在阳光的照射下形成阴影产生美感。

The plaza design sources its ideas from Manchu culture and mineral products, two major features of Fushun city. Metallic plates and finely designed modular window decoration texture perfectly harmonize classic and modern features. The shade under sunshine formed by projection of perforated metal panels creates a sense of beauty.

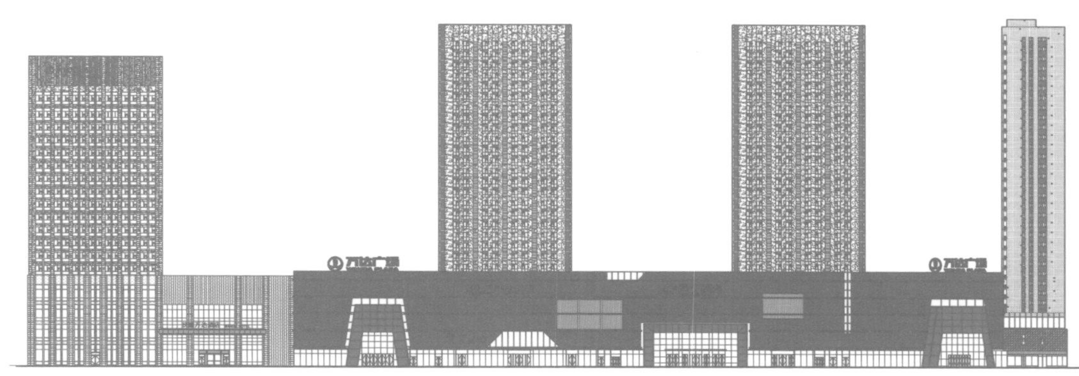

4

5

3 广场外立面
4 广场立面图
5 广场主入口

WANDAPLAZA
10月25日
万达广场
WANDA PLAZA
WANDAPLAZA

6b

6 广场立面特写

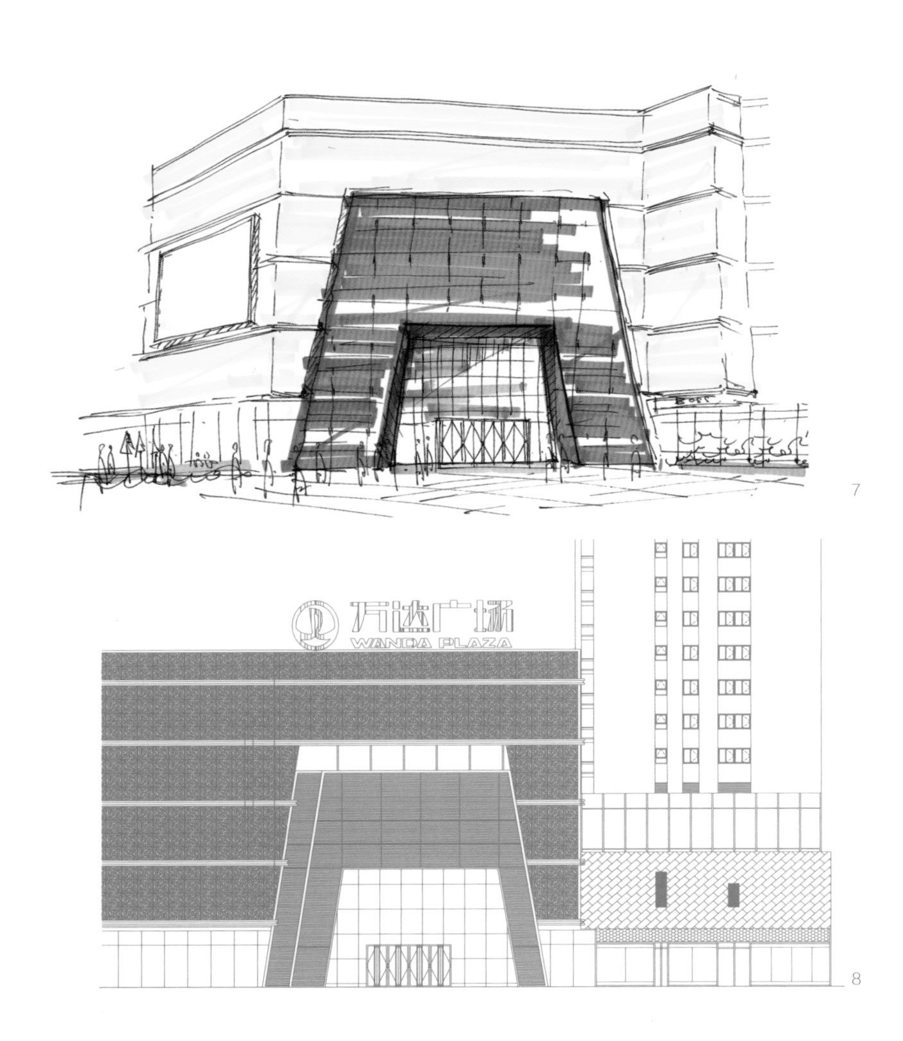
万达广场
WANDA PLAZA

7

8

万达广场
WANDA PLAZA
尝试 玩美
挑战
享受...
a world of beauty
SEPHORA

9

10

大商业的主入口设计，以抚顺盛产的稀有矿物“血珀”为设计主题，在保持了整体感和大气感的同时，利用丝网印刷玻璃形成大体量的门头吸引人流进入商场，成为强烈的视觉中心，尺度上改变原有的比较冷漠与单调的界面，增加材料质感与种类的统一性，使市民在万达广场内的空间体验得到改善。

The main entrance design for the large commercial center is themed by "blood amber", a rare but abundant mineral in Fushun. While maintaining integrity and grandness, a large-sized gate made of screen printed glasses is designed, which attracts pedestrian flow into the store, forms a strong visual center and changes the original cold and monotonous interface by integrating material texture and category so as to improve spatial experience of citizen in Wanda Plaza.

7 广场入口设计手稿
8 广场入口立面图
9 一号入口
10 百货入口

INTERIOR OF PLAZA
广场内装

抚顺是“启运之地，满乡故里”，原为大清龙脉所在，采光顶以“龙行天下”为母题，广场室内设计元素以龙鳞为根基，整条室内步行街宛如“龙戏二珠”。夕阳西下的暮色之中，抚顺万达广场像一条卧龙般潜伏于浑河之畔。

Fushun is known as "the place to open one's fortune and the hometown of Manchu", and enjoys a strategic location in the Qing Dynasty. Daylighting roof design is themed by "dragon traveling around the world"; interior design of the indoor pedestrian street, based on dragon skin element, creates an image that "dragon playing with two pearls". The plaza looks like a lying dragon on the bank of Hun River under twilight.

龙游江海，波光粼粼，这是在主入口设计上所要突出的，延续一贯的设计元素——龙鳞、造型、灯光只为突出“波光粼粼”的视觉感，千变万化，一切让人流连忘返。

The image of dragon wandering in glittering river water is what the main entrance design highlights and a design element to be adopted. The shape and light design are only aimed to stress such visual impression of glittering water with variations so as to attract people to stay.

12

13

11 入口门厅天花
12 入口门厅侧壁
13 入口天花

14

15

椭圆中庭是整个空间的重中之重，中庭侧裙整个形体好似云雾中腾飞的蛟龙，若隐若现；侧裙留白部分似云似雾，侧裙上的菱形就如同龙身的龙鳞；而点缀的不锈钢如同阳光下闪烁的龙鳞，起到画龙点睛的作用，且采光顶钢结构和观光梯的菱形也体现了龙鳞元素，丰富了整个空间。

Oval atrium is priority among priorities of the whole space, side plates are arranged in a way like flood dragon soaring among clouds and mists, partly hidden and partly visible; blank part of the side plates sometime appear like cloud and sometimes like mist, and rhombuses on side plates are like dragon skin; scattered stainless steel is the finishing touch, looking like dragon skin sparkling under sunshine. Rhombic steel structure on daylighting roof and rhombic part of panoramic lift also embody dragon skin element, thus enriching the whole space.

14 椭圆中庭设计手稿
15 椭圆中庭
16 椭圆中庭采光顶

开业大请客
美食全场
5
折起
éifini
JDING
AMASS

gelato & waffle
caffé bene
VANS
Paul Frank

18

圆中庭两边的侧裙板以龙鳞的花纹做出了肌理且富有雕塑感的设计造型，在灯光下的作用下，就好像一条腾飞的龙浑身散发着光芒，两边走廊的天花造型不再是柔软有弧度的线条，而是转变成有菱角的刚硬线条，与椭圆中庭的天花相对应。观光电梯一改往常用的龙鳞元素，而将其演变成矩形加以错落有致地排列，取得出神入化的效果。

Side plates on both sides of the circular atrium are designed in dragon skin patterns and a sculpted shape, which looks like a soaring dragon emitting brilliant rays under the light effect. Suspended ceilings on both sides of corridors are in rigid lines with edges instead of flexible arched lines, which echo with the suspended ceiling of the oval atrium. Instead of adopting commonly used dragon element in panoramic lift design, rectangular patterns are arranged in a well-proportioned manner to realize a superb effect.

19

17 圆中庭采光顶
18 圆中庭
19 圆中庭设计手稿

20

长街的设计未做过多的装饰，天花造型趋向于简单化，地面拼花跟随着天花，两者相得益彰。长街的侧裙板加以简化处理，以折线和几何块面为设计元素的表现形式，增加了空间的时尚感和现代感，而如同飞龙般蜿蜒曲折，使空间更加灵动，丰富了人们的视觉空间。廊桥作为两侧长街的连桥，两侧侧板采用不锈钢材质做出具有凹凸纹理的装饰效果，雕塑性强，远看就像立体的龙鳞展示在眼前，使两边侧裙板看起来层次更加丰富。

20 室内步行街
21 室内步行街天窗
22 连桥转角
23 自动扶梯

21

22

The long gallery design is featured by less decoration and suspended ceiling is in a simple shape. Floor patterns and suspended ceiling patterns bring out the best in each other. Side plates for the long gallery are simplified, adopting broken lines and geometrical blocks as design elements to create a fashionable and modern space which looks like a winding dragon, making the space dynamic and enriching people's visual space. Gallery bridge connects the long galleries on both sides, and stainless steel side plates is designed with concave and convex texture to show a highly sculpted effect, looking like a three-dimensional dragon skin when stand afar and enriching sense of depth of side plates on both sides.

23

24

LANDSCAPE OF PLAZA
广场景观

抚顺万达大商业项目的景观设计，在空间上营造出与建筑和周边环境浑然一体。两大文化主题的巧妙融入，又鲜明地表达了抚顺这个城市的时代人文气质。抚顺被称为“雷锋之城”，设计通过卡通人物形象作为载体，其肢体语言与物体形象配合的抽象表达，体现了对新时代新生活的理解和感悟，生动地展现了新时代下的雷锋精神，传递社会正能量。

“满清文化”的表现，运用了更为直接的现代极简主义处理手法——从诸如宫女头饰、满清服饰中的梯形元素、器具中的折线元素中提炼，除了呼应商业建筑和广场立面的“满清文化”元素，同时进行提炼简化和变化对接，在景观的限定和划分功能空间、地面铺装和材质等细节中完美呈现。

Landscape design for Fushun Wanda large commercial center project creates a space where the architecture is perfectly integrated into its surrounding environment. Two cultural themes are skillfully integrated into the landscape design, vividly conveying humanity temperament of Fushun in modern times. Fushun is called as "the city of Leifeng", so the project landscape design takes cartoon figures as its carrier, combined with abstract expression via body language and subject image, expressing people's understanding and perception on the new era and new life and vividly showcasing Leifeng's spirit in the new era and passing on positive energy to the whole society.

Direct modern minimalism treatment methods, including extracting ideas from trapezoid elements such as headwears of maids in imperial palace and Manchu costume as well as broken line element in appliance, are used to express "Manchu Culture". Besides corresponding to commercial building and "Manchu Culture" elements on the plaza façade, these elements are extracted, simplified and docked to perfectly showcase "Manchu Culture" through such details as landscape restriction, functional space division, floor pavement and materials selection.

24 “雷锋”主题卡通雕塑
25 广场座椅平面图
26 广场景观
27 广场座椅

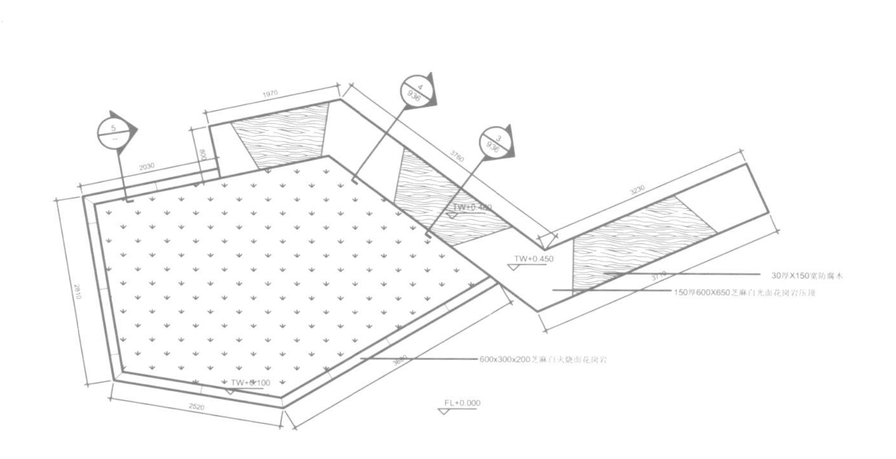

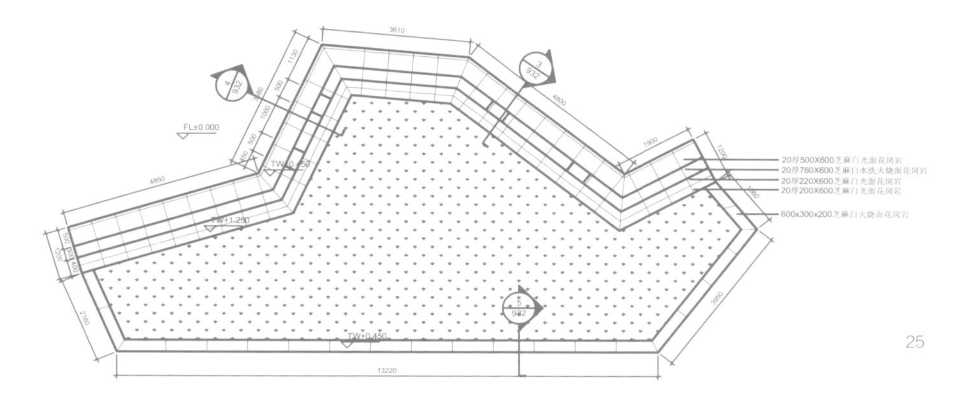

25

26

27

NIGHTSCAPE OF PLAZA
广场夜景

以抚顺“龙兴之地”传统文化为主题，通过灯光对“满洲中式窗花”纹路结构的诠释，深层次挖掘当地人文及城市底蕴，光影变化在营造浓厚商业氛围同时将满清文化、雷锋之城及万达精神等元素与现代建筑设计巧妙融合充分表现建筑与城市内在的紧密联系。俊朗的白色线条勾勒塔楼轮廓，璀璨的十字星光富有动感，暗示着雷锋精神的永恒。工业液体的流动到固化的过程的抽象语言，用灯光的理念在夜间诠释，表达着这个城市经济发展的主支柱。光学幕墙在大商业主入口的运用，构成了丰富迷离的灯光层次，动感深邃、富有强烈的视觉冲击力，预示着未来。

Themed by the traditional culture of Fushun as "the Land of Dragon Rising", the nightscape design attempts to explore local human and urban deposits via lights design which interprets a "Manchu Chinese-style window decoration" grain. Variations in light and shadow skillfully integrate such elements as Manchu Culture, the City of Leifeng and Wanda Spirit into modern architectural design while creating a strong commercial atmosphere, thus fully demonstrating close and inherent relationship between architecture and the city. Handsome white strokes draw the outline of tower building, resplendent cross starlight is dynamic, implying eternity of Leifeng's Spirit. Abstract language of industrial liquid from flowing to solidification is interpreted by lighting design at night, expressing main backbone of urban economic growth. Optical curtain wall at the main entrance of the large commercial center constitutes rich and blurred light layers, which is dynamic, profound, with strong visual impact and indicating the future.

28 广场鸟瞰夜景

万达广场
WANDA PLAZA
万达百货 万达影城 大歌星 KTV
乐天玛特 10月25日盛大开业
临街
投资首选
精装
SOHO
即买即赚
临河
火爆热销

29

29 广场外主入口
30 广场外立面
31 百货门头

30

THE OUTDOOR PEDESTRIAN STREET
室外步行街

32 室外步行街设计手稿
33 室外步行街外立面
34 室外步行街立面图
35 室外步行街外立面夜景
36 室外步行街门头

32

33

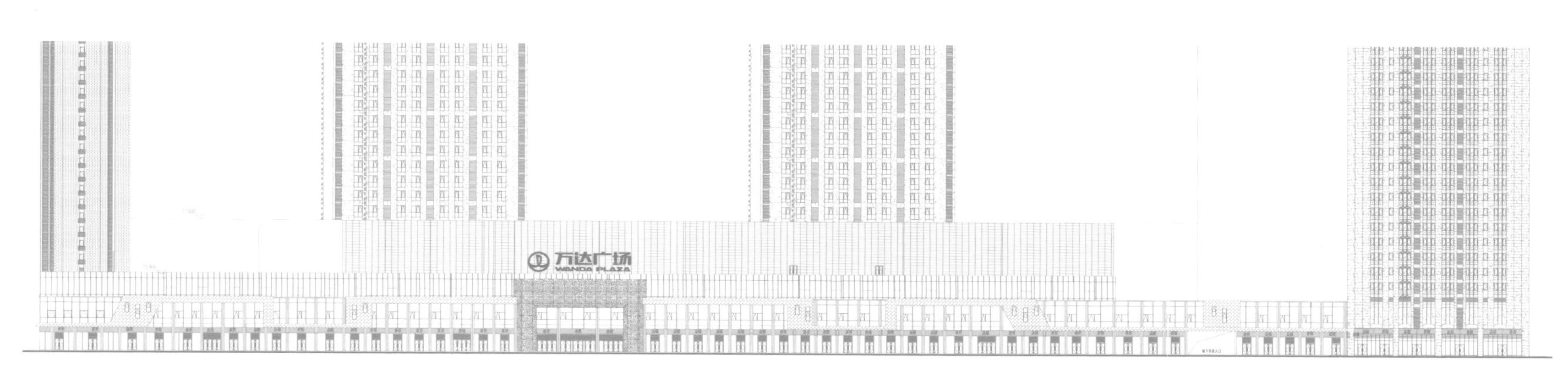
万达广场
WANDA PLAZA

34

10月25日
盛大开业
WANDA PLAZA
Coffee

35a

PEACE
2013
10.25

35b

万达广场
2013
10.25
来店礼

36

FACADE OF HOTEL
酒店外装

酒店外立面设计上采用大气端庄、简洁明快的手法。重视与城市环境的互动关系，使建筑与环境充分协调又独具个性。整体立面吸取了在庄重的古典建筑中常用的三段式设计手法。塔楼顶部采用密集壁柱形式，与主体相辅相成，形成方案的视觉中心。强烈的虚实对比，既传统又时尚。塔楼中段采用简洁的米黄色仿石涂料加深窗洞的形式，增加了光影变化，立面整体感强，简洁而有力度。裙楼采用四层通高的落地大窗与厚重石材包裹，简洁大气；所有幕墙玻璃分隔立挺均与塔楼窗洞及玻璃分格对应，逻辑关系清晰，对位严谨，与入口巨型雨棚所呼应，彰显奢华和尊贵的品质。

38

Grand while concise hotel façade design attempts to value its interactive relationship with urban environment, fully coordinating the building and environment while keeping its individuality. The overall facade design adopts three-section design method, which is commonly used in solemn classic architectural design. The top part of the tower building is arranged with densely allocated pilasters to supplement the main body and becomes a visual center. Strong virtuality-reality contrast is traditional but fashionable. Simple off-white stone paint plus deep window opening at the central section of the tower building enriches shadow variations, with a strong sense of integrity, concise but powerful. Podium building is designed with four-storey and full-height French window, wrapped with thick stone materials, concise and grand; glass divisions of curtain walls correspond to window openings and glass divisions of the tower building, with clear logic relationship and rigorous alignment, corresponding to the giant canopy at the entrance and manifesting luxurious and noble quality of the hotel.

37

37 酒店立面图
38 酒店门头
39 酒店外立面

WANDA REALM
WANDA REALM
大型超市
乐天
玛特

40

LANDSCAPE OF HOTEL
酒店景观

40 酒店景观绿化
41 酒店景观铺装
42 景观亭立面图
43 景观亭
44 景观走廊

41

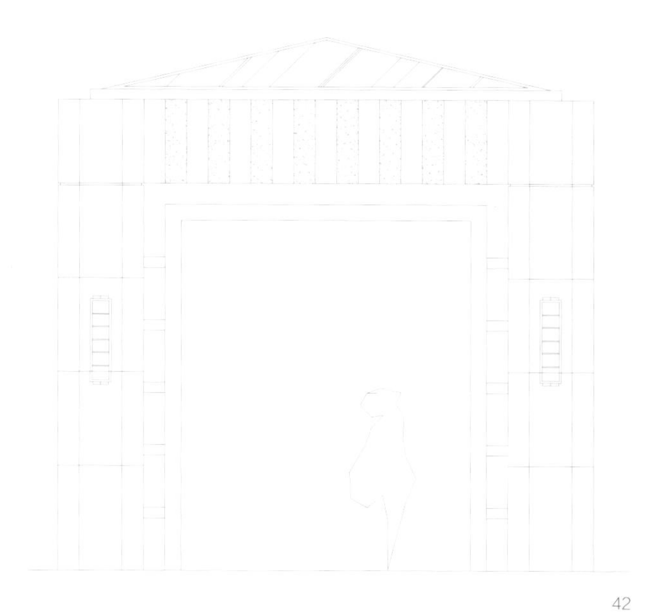
42

43

44

NIGHTSCAPE OF HOTEL
酒店夜景

酒店夜景照明利用立面洗墙及轮廓灯等简洁的照明手法表现建筑层次，营造温馨高档奢华的氛围，在舒适宜人的光环境中提升酒店整体的品位和层次。大尺度壁灯在连续立柱结构上形成了强烈的序列感。楼体四角的轮廓灯追逐呈灰度动态渐变勾勒出楼体的夜间形态。入口雨棚采用下照式筒灯与LED隐藏灯带相结合的做法，充分表现雨棚顶部结构及建筑材料肌理的特性，在满足入口照度要求的同时增强酒店入口位置感及重要性。

Nightscape lighting design for the hotel attempts to express architectural levels via facade flood lights, contour lights and other concise lighting methods, thus creating a comfortable and luxurious atmosphere and elevating overall taste and level of the hotel in a pleasant light environment. Large-sized wall lamps on continuous stand column structure form a strong sense of order. Contour lights on four corners of the building adopt gray scale gradient in a dynamic way so as to draw the outline of the building at night. Canopy at the entrance adopts both down light and LED concealed light band, fully showcasing characteristics of caisson structure at the canopy top and building material texture, and enhancing the sense of position and importance of the hotel entrance while meeting illuminance requirements.

45 酒店入口喷泉
46 酒店外立面

WANDA REALM
抚顺万达嘉华酒店
乐天
玛特

NINGBO YUYAO WANDA PLAZA
宁波余姚万达广场

开业时间 2013 / 11 / 01
建设地点 浙江 / 宁波
占地面积 11.45 公顷
建筑面积 35 万平方米

OPENED ON NOVEMBER 1 / 2013
LOCATION NINGBO / ZHEJIANG PROVINCE
LAND AREA 11.45 HECTARES
FLOOR AREA 350,000 m^2

OVERVIEW OF PLAZA
广场概述

余姚万达广场位于浙江省余姚市兰江区，四明西路以南，开丰路以西。余姚万达广场总建筑面积 35 万平方米，由购物中心、室外步行街、精装住宅等构成，其中购物中心 17 万平方米（地上 9 万平方米，地下 8 万平方米），室外步行街商铺 3 万平方米，公寓 1 万平方米，商务酒店 1 万平方米，精装住宅 13 万平方米。

Ningbo Yuyao Wanda Plaza is located in Lanjiang District, Yuyao of Zhejiang Province, to the south of West Siming Road and to the west of Kaifeng Road. The plaza has a gross floor area of 350,000 m^2 and is composed of shopping center, exterior pedestrian street and finely decorated residence. The shopping center is 170,000 m^2 in area (90,000 m^2 for aboveground part and 80,000 m^2 for underground part), the exterior pedestrian street store has a floor area of 30,000 m^2, the apartment 10,000 m^2, commercial hotel 10,000 m^2 and finely decorated residence 130,000 m^2.

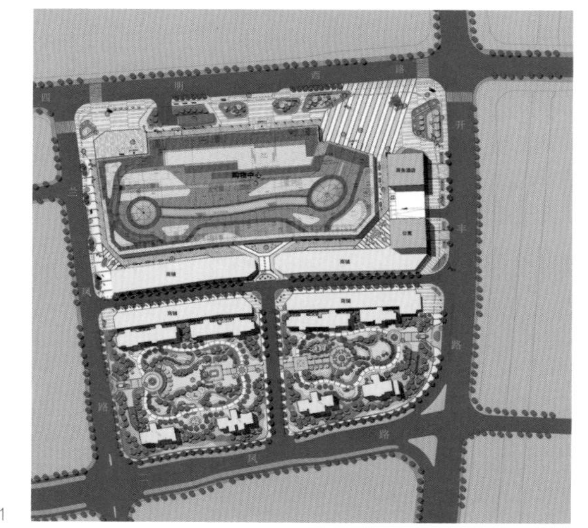

1

1 广场总平面图
2 广场鸟瞰图
3 广场外立面

4

FACADE OF PLAZA
广场外装

余姚是河姆渡文化的发源地，姚江水孕育着世代勤劳智慧的余姚人。水是最能体现余姚当地文化的载体，立面构思以此为出发点，并着力表现。

建筑立面强调优美有节奏的韵律变化，主体立面犹如弧形的波浪，让人印象深刻。内外表皮采用视错觉的设计手法，不同的角度，呈现不同的视觉美感。通过简洁流畅的构图，白色基调配以彩色元素，传递时尚、动感，有活力的购物中心形象。幕墙外表皮由 156 根弧形杆件排列而成，运用参数化设计手段来控制每根杆件的长短，形成富于韵律变化的曲线构图，远观达到波浪起伏的视觉效果。

4 外立面特写
5 百货外立面
6 广场立面图

5

Yuyao is the cradle of Hemudu Culture, and Yao River is the mother river of generations of industrious and intelligent Yuyao people. Water is the best carrier of local culture, so the Plaza elevation design concept originates from water and focuses on water element expression.

The building façade design highlights graceful and rhythmic variations, and main elevation appears like arc-shaped waves, which is impressive. Interior and exterior surfaces adopt illusional design method to showcase different visual aesthetic perception from different aspects. Concise and smooth composition, white tone and color elements, all these create a fashionable, dynamic and vigorous image of shopping center. External surface of curtain wall is made of 156 arc rods, with length of each rod controlled by means of parametric, thus forming a rhythmic curve and a lumpy visual effect when stand afar.

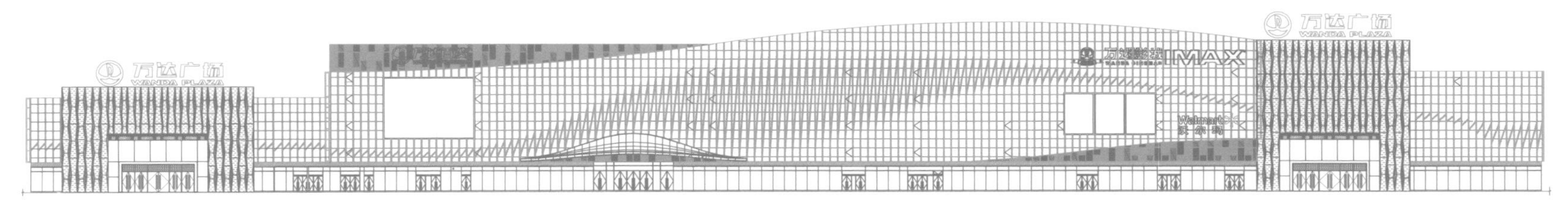

6

7 广场雨棚

8

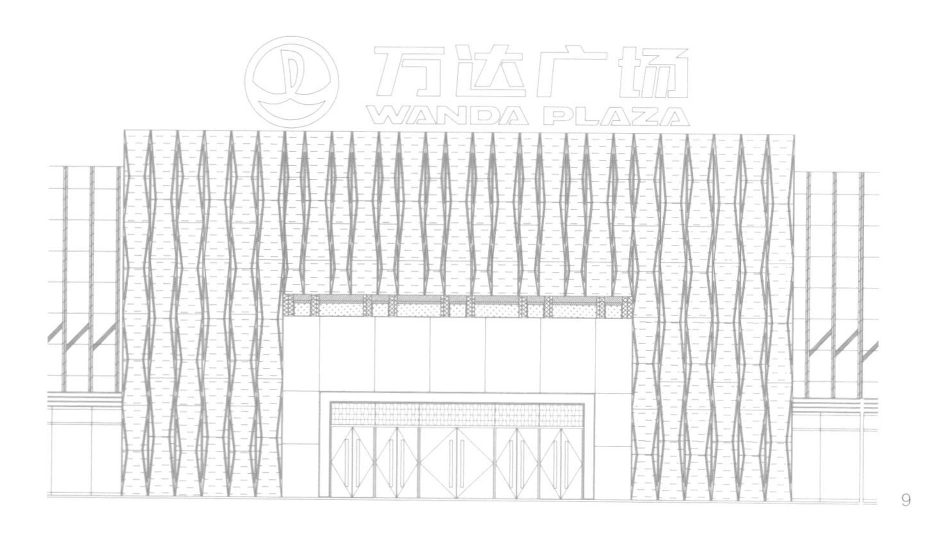

9

10

玻璃门头主入口与主体虚实对比强烈，精致的“梯形”玻璃单元错动排列，富有节奏，给人以强烈的视觉冲击力。二层采用内透设计，透过玻璃可直接看到购物中心内部。内透处理增强了商业的展示效果。双曲面玻璃雨棚特色鲜明，造型与“弧形”立面的紧密结合，同时凸显了百货主入口的形象。

Main entrance of glass gate and main body constitute a strong virtual-real comparison, and delicate "trapezoid" glass unit arranged in a staggered way to generate a sense of rhythm and strong visual impact. On the first floor, the interior of the shopping center can be seen through glasses, thus enhancing demonstration effect of the commercial center. Distinctive dual-curved surface glass canopy is in a shape closely linked to "arc" elevation, highlighting the image of the main entrance of the department store.

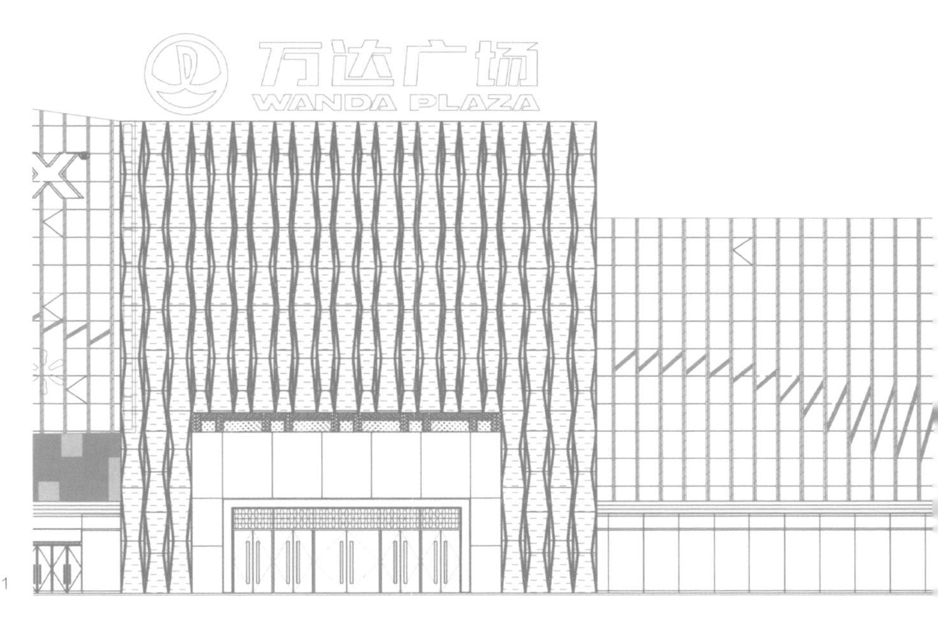

11

8 广场一号入口
9 一号入口立面图
10 广场二号入口
11 二号入口立面图

12

12 圆中庭
13 河姆渡时期文物
14 圆中庭地面花纹

13

INTERIOR OF PLAZA
广场内装

河姆渡文化是中国长江流域下游地区古老而多姿的新石器文化，本案设计概念从河姆渡文化出发，选取河姆渡文化中人们生产生活中所使用的纹样，并与现代设计语言和处理手法相结合，展示既现代又具有当地文化特色的内装风格。

河姆渡时期最典型的交通工具是扁舟，人们通过这种交通工具进行生产生活。圆中庭侧帮叶子形的造型是对一叶扁舟这种概念的高度提炼，地面的水波纹，寓意舟在河水中穿行。地面花纹样式也是参考了该时期的陶瓷纹样进行的再次设计。

Hemudu Culture is an ancient but colorful Neolithic culture for people living at the lower reaches of the Yangtze River. The Plaza design selects patterns used in production and daily life of people in Hemudu Culture and demonstrates them via modern design language and treatment method, thus showcasing the interior finishing which is modern and of local cultural characteristics.

The most typical means of transportation during the Hemudu Period was small boat, an essential tool for production and daily life. The circular atrium is designed with a leaf-shaped lateral wall by highly refining the concept of small boat; water ripple patterns on floor implies that small boat is passing through river; floor patterns are a kind of re-design by referring to ceramic patterns during the period.

14

15

整个中厅的空间情绪展现了对河姆渡文化的尊重与敬仰，观光梯从设计形态上是对瓷器文化与河姆渡文化的理解与传承，地面花纹样式也是参考了该时期的陶瓷纹样进行的再次设计。

Spatial mood of the whole atrium demonstrates a respect and admiration to Hemudu Culture; panoramic lift design form reflects understanding and passing on of ceramic culture and Hemudu Culture; floor patterns are a kind of re-design by referring to ceramic patterns during the period.

16

15 河姆渡时期文物
16 椭圆中庭地面花纹
17 椭圆中庭

11月1日盛大开业
GRAND OPEN
watsons
GUCCI
MIX & MORE
万达百货
dax
UNI QLO

18

整个空间侧帮使用了河姆渡文化的纹样，用洗练的设计语言将瓷器纹样用现代的手法和材质进行表现，传承了河姆渡文化精神。采光顶分格采用中国传统纹样设计。为了让廊桥在整个长街空间里显得轻薄，将其设计成一个半包围的发光盒子，既烘托出商业氛围，又在视觉上减弱了桥在长街空间中的比重；形式上也是延续了河姆渡文化的纹样。

The lateral wall, designed with patterns used in Hemudu Culture. Ceramic patterns are presented via sophisticated means and modern methods & materials, which is a kind of passing on the Hemudu Culture spirit. Traditional patterns are adopted in daylighting roof partition. The gallery bridge is designed as a semi-encircled glowing box to make it light and thin in the whole long gallery space, which not only highlights commercial atmosphere but also visually weakens the proportion of the bridge in the long gallery space. Besides, the gallery bridge adopts patterns used in Hemudu Culture.

19

20

18 室内步行街
19 连桥
20 自动扶梯

ONE STORE, ONE STYLE
一店一色

21a

21b

21c

22

21 特色店面
22 商业侧裙板

23

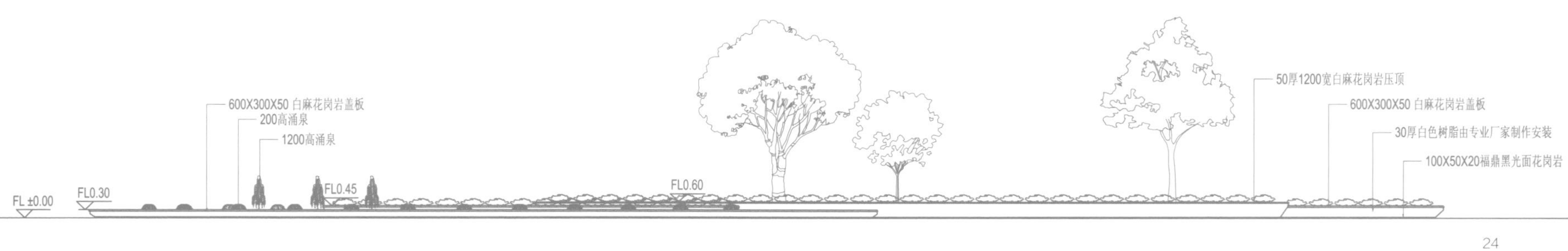

24

25

23 广场绿化
24 水景立面图
25 花坛
26 花坛平面图
27 广场地铺

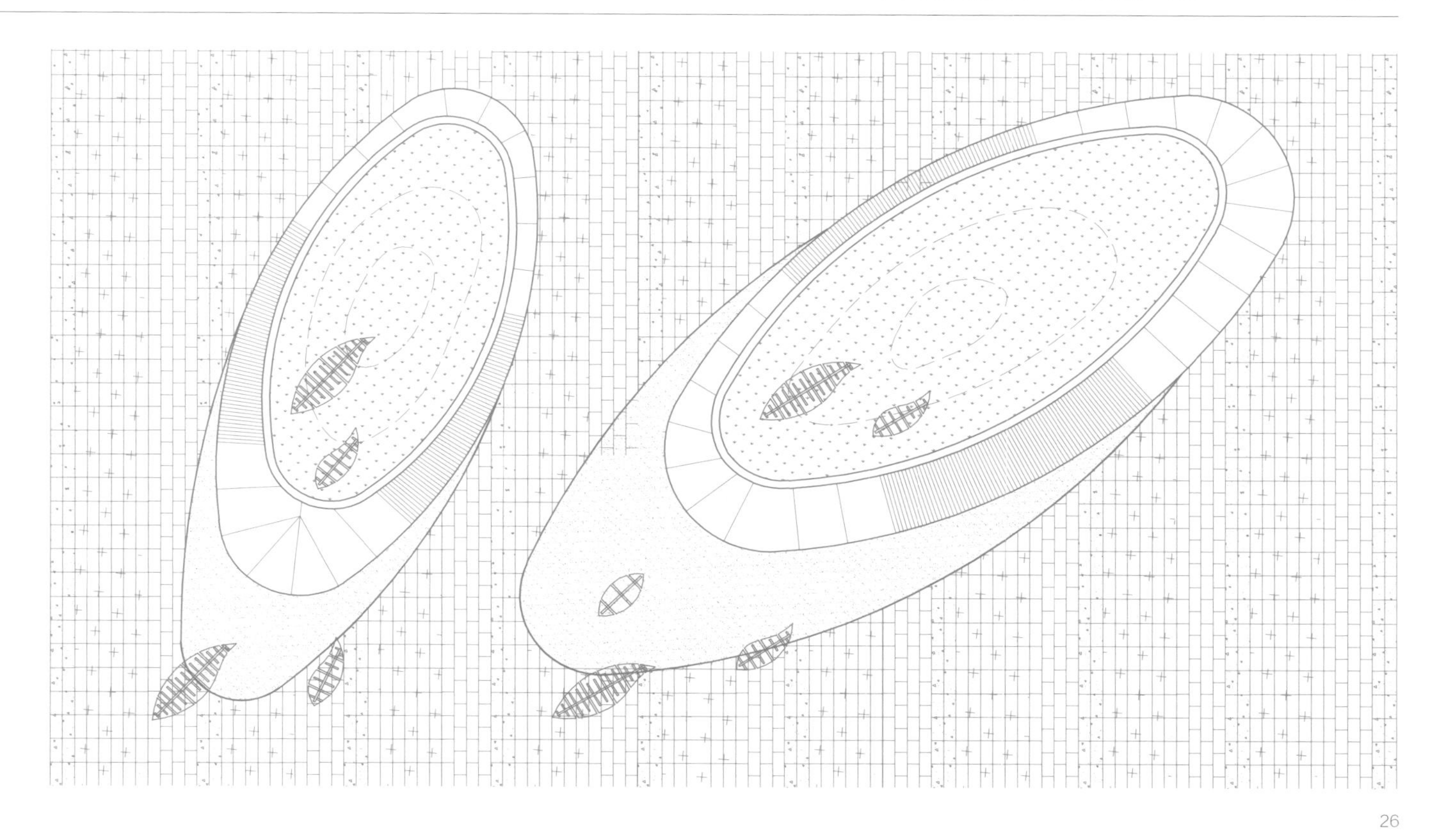

26

27

LANDSCAPE OF PLAZA
广场景观

景观采用余姚市市花“玉兰”为主题，椭圆形的植物岛及周围铺装犹如一片片玉兰花瓣般散落在广场上——大小有别，错落有致，既简洁又现代、大气；既给人以开敞的商业空间享受，同时又不缺细节展现。

Themed by "Yulan magnolia", the city flower of Yuyao, an oval plant island is arranged, with its surrounding pavement looking like Yulan magnolia petals scattering on the plaza: large and small, high and low, concise while modern and grand, offering people a spacious commercial space enjoyment while demonstrating details.

NIGHTSCAPE OF PLAZA
广场夜景

大商业主立面LED灯条借用建筑的竖向杆件，构件的正面、背面设计为两层灯具，正面作为灯具直接表现界质，背面灯具借用建筑立面墙体作为灯光二次反射界质。灯光效果分层显示，实现了在同一平面上，表现出多层的灯光效果，层层可以独立表现，也可以互相呼应，形成立体的灯光效果。整个画面相对单层显示来说，画面更饱满、层次更丰富、表现内容更多样，形成独特的灯光效果。

LED light bars on main elevation of large commercial center borrow vertical members on the building. The front and back sides of the members are designed with two-layer lighting fixtures, with the front side directly showcasing lighting fixture while the back side lighting fixture serves as secondary reflector by borrowing building elevation wall. Compared to single layer display, the picture as a whole becomes more mature, with rich layers, diverse contents and unique light effect.

28

29

28 广场鸟瞰夜景
29 广场外立面夜景

 广场外立面夜景

万达影城
MAX

32

33b

31 雨棚夜景
32 二号入口夜景
33 外立面夜景图案

OUTDOOR PEDESTRIAN STREET
室外步行街

室外步行街设计了一系列凸出的飘板，在色彩和形态方面都有不同的处理。飘板外表面采用三种颜色间隔变化，形态上高低变化，造型犹如折纸一般连接整个商街，统一连续。鲜明的色彩、变化的造型，使在步行街购物的顾客充分感受到活跃的商业氛围。

The exterior pedestrian street design is featured by a series of protruded floating plates in different colors and shapes. The exterior surfaces of such plates adopt three colors changing at an interval, high and low in shapes, connecting the whole street like folded paper to maintain unity and continuity. Vibrant colors and changing shapes offer customers a dynamic commercial atmosphere.

35

34 室外步行街入口
35 室外步行街雕塑

36

37

36 室外步行街外立面
37 室外步行街转角
38 室外步行街夜景
39 景观小品

38

39

XI'AN DAMING PALACE WANDA PLAZA
西安大明宫万达广场

开业时间 2013 / 11 / 22
建设地点 陕西 / 西安
占地面积 10.52 公顷
建筑面积 64.68 万平方米

OPENED ON NOVEMBER 22 / 2013
LOCATION XI'AN / SHANXI PROVINCE
LAND AREA 10.52 HECTARES
FLOOR AREA 646,800 m^2

OVERVIEW OF PLAZA
广场概述

西安大明宫万达广场位于西安市未央区，毗邻著名的大明宫遗址公园。项目总占地 10.52 公顷，总建筑面积 64.68 万平方米，集大型商业购物、时尚步行街、精装 SOHO 和 5A 级商务办公中心为一体，成为引领古都西安时代商业潮流的国际级购物中心。

Xi'An Daming Palace Wanda Plaza is located in Weiyang District of Xi'an, adjacent to the famous Daming Palace National Heritage Park. Covering a land area of 10.52 hectares and a gross floor area of 646,800 m^2, the Plaza is a multi-functional space with such functions as large commercial shopping, fashionable pedestrian street, finely decorated SOHO and 5A grade commerce and office center, and a leading international shopping center in the ancient city of Xi'an.

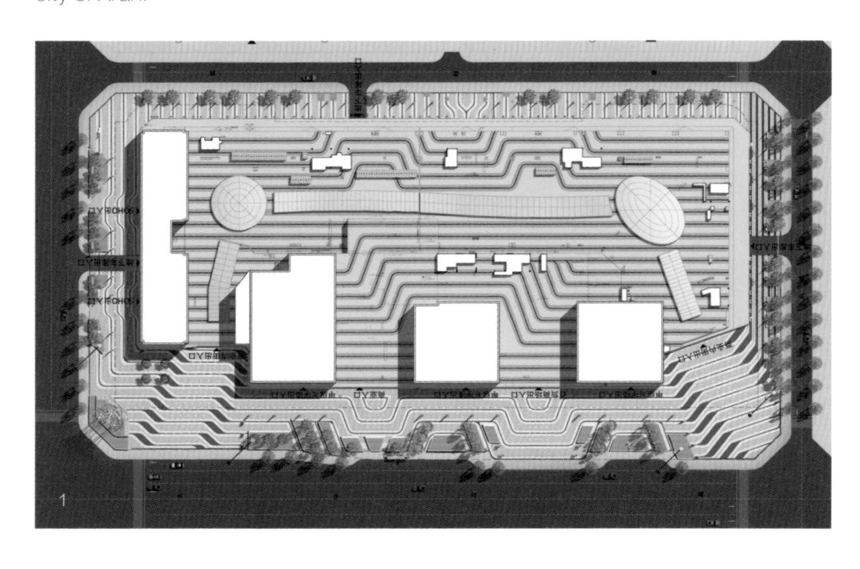

1 广场总平面图
2 广场外立面

2

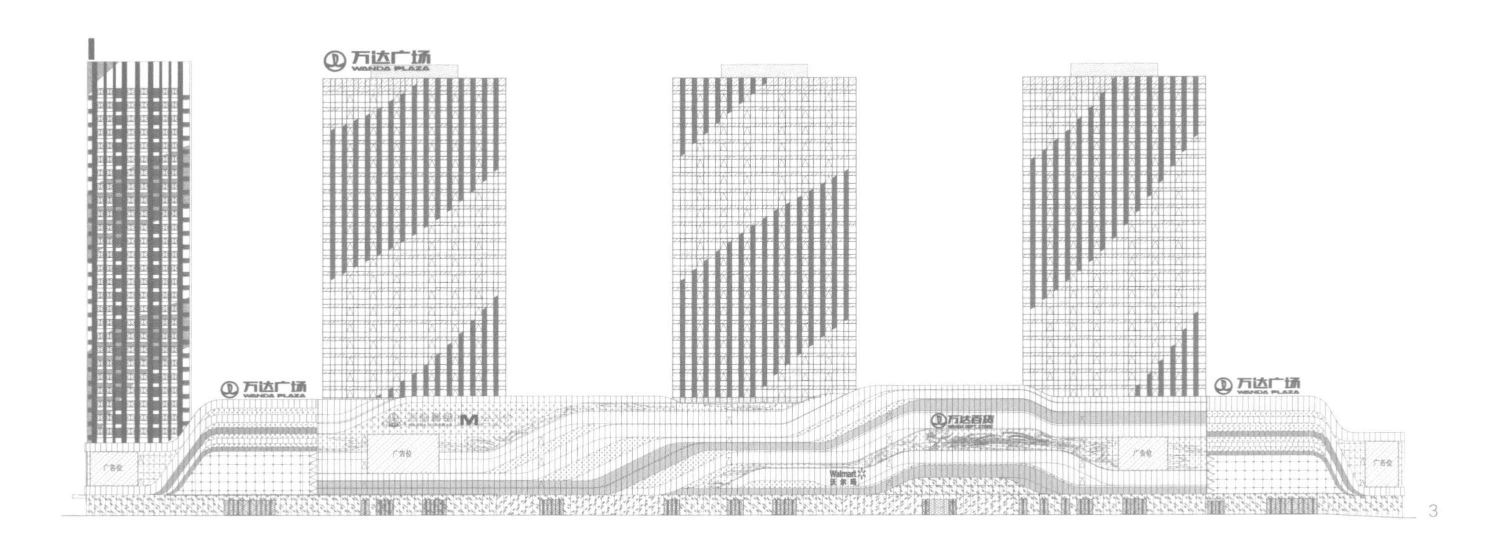

3

5

4

FACADE OF PLAZA
广场外装

整体造型上提取“飞天飘带”概念，通过抽象的语言重塑，流畅的水平线条与曲线相结合，贯穿整体，造型简洁，大气时尚，以现代的造型语素，隐喻了深刻的历史文化元素。

In overall shape of the Plaza is designed by extracting the concept of "ribbon flying in the sky", abstract language remodeling, combination of smooth horizontal lines and curves, creating an integral, concise in shape, grand and fashionable space, and implying profound historic and cultural elements with modern shape language.

3 广场立面图
4 外立面正视图
5 立面局部特写

门头处理，抽离中国古代建筑的窗棂形式作为母体，用金色金属材料，布置成气势宏大的门头肌理；在玻璃的处理上，使用彩釉工艺，用金色点阵形成层次丰富的玻璃幕墙，创造了特色鲜明的现代商业主入口形象。

The gate design element is extracted from window lattice used in ancient Chinese architecture, adopting golden metallic materials to showcase grand texture of the gate; color ceramic glaze glasses are arranged in golden matrix so as to generate glass curtain wall of rich layers, thus creating a distinctive image of main entrance of modern commercial center.

6

7

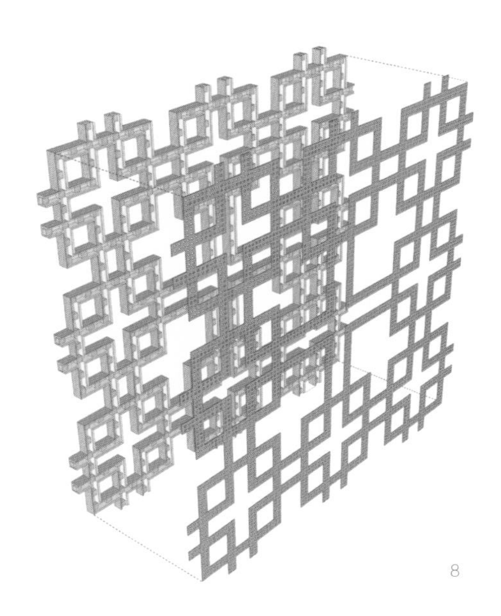
8

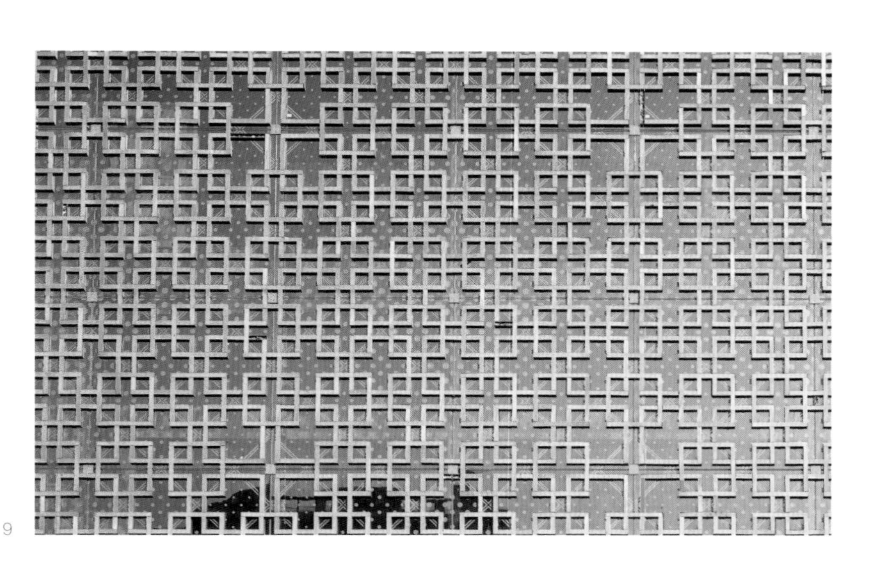
9

10

6 广场一号入口
7 广场二号入口
8 门头窗花构造
9 门头窗花图案
10 窗花局部特写

11

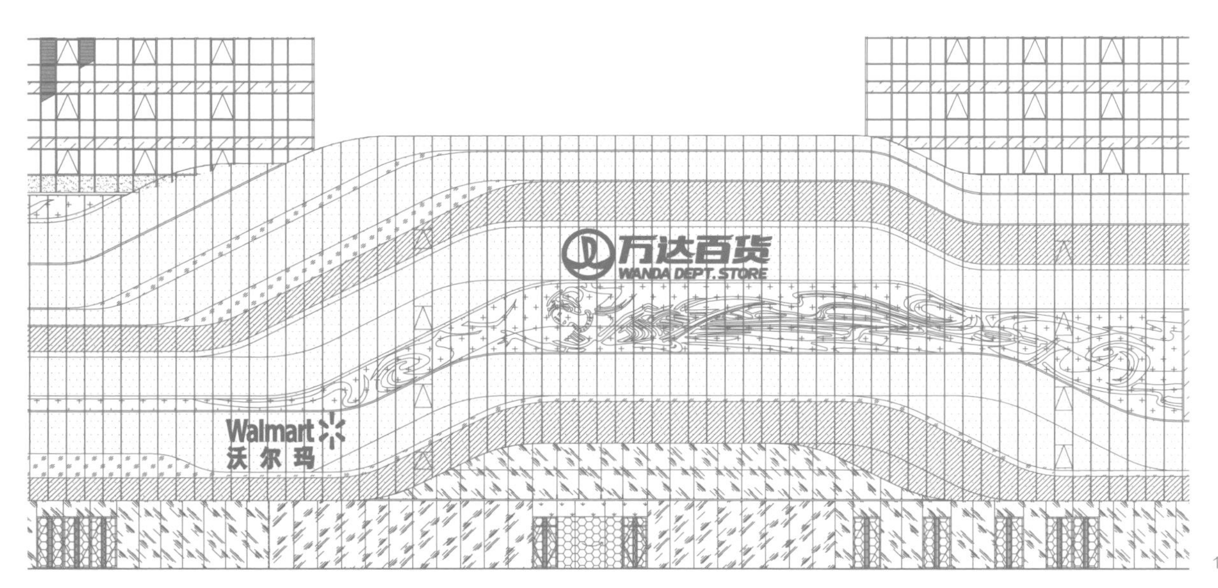

12

11 百货外装立面
12 百货门头立面图
13 百货门头局部特写
14 百货入口

立面细部处理，与主体的抽象手法相结合，通过材料颜色和形式的变化——天蓝色铝板结合穿孔处理形成具象的飞天造型，生动地、多层面地强调“飞天”立面主题。

Detail treatment of elevations is based on abstract treatment of the main body. Variations of colors and shapes of materials-sky blue perforated aluminum plates presenting a concrete flying shape, which vividly and multi-dimensionally stresses the elevation theme of "flying to the sky".

13

14

INTERIOR OF PLAZA
广场内装

椭圆中庭无疑把“行云流水，千载古城”的设计理念贯穿到极致。侧裙采用双层结构，一气呵成的结构曲线由侧裙延伸到天花，材料上采用珍珠白色的GRG成品；同时搭配内部透光的效果，再加上结构双层藏光形成细腻的空间构成艺术效果，远处看犹如两条流动的曲线在空中互相变化。椭圆中庭的整个空间就围绕着这两条贯穿整个空间的流线展开，再加上椭圆观光电梯的镀膜玻璃的曲线纹理以及精致的图案一直延伸到天和地，使整个空间舒展大气，又不乏细节的艺术点缀，让人流连忘返。

Oval atrium undoubtedly well demonstrates the design concept of "floating clouds and flowing water, an ancient city with a thousand year's history". Double-layer side skirt structure made of pearl white GRG finished product stretches to suspended ceiling, which is coherent; meanwhile, interior permeable light effect and inserted double-layer light jointly form a fine artistic effect, which seems like two flowing curves changing alternatively in the sky when looking afar. The whole oval atrium space is designed by centering on the two streamlines, oval panoramic lift with curve texture on coated glass and delicate patterns which extends to the sky and ground, creating a stretched and grand space full of artistic embellishment, thus attracting people to linger on.

15

16

15 椭圆中庭地面俯视

16 椭圆中庭

SELECTED
JUSTUS
Desires

17

18

圆中庭同样使用了"模数化"设计，与椭圆中庭相比较，侧裙的曲线显得更加硬朗，但保留了造型一直延伸到天花的特点。设计将圆中庭侧裙图案的序列、大小、颜色加以变化，进而强化了流动性；并采用浅暖灰色的金属面漆把整个造型突出，结合藏光的形式，让设计更有生命力。

The circular atrium also adopts "modular" design, but side skirt curves of the circular atrium are harder than that on the oval atrium, and the shape of the circular atrium also extends until the suspended ceiling. Variations in sequence, size and color of side skirt patterns on the circular atrium further enhances mobility; the overall shape is highlighted by metallic finishing coat in light warm gray color, combined with inserted light, making the design more vigorous.

19

17 圆中庭采光顶
18 圆中庭侧裙板
19 圆中庭

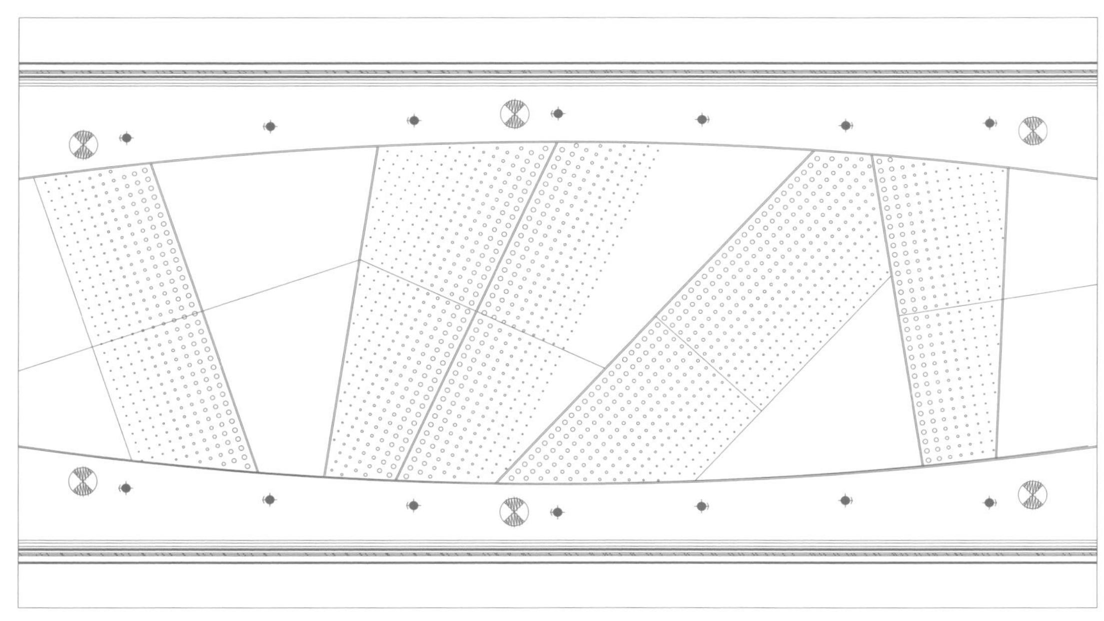

20

长街波浪造型与每个过桥有直接的呼应关系，匠心独运的设计使得每个过桥都在波浪的最低点处。在此基础上，侧裙板的波浪起伏与流线图案又在其中有所变化。长街的“棱形过桥”无疑是整个长街的画龙点睛之处，过桥的造型犹如半个“棱形”轻盈地镶嵌在长街中，过桥的重点偏向一端，使得原先笨拙的过桥富有活力。结合环形的藏光，令过桥更轻盈更富于现代感。传统的文化符合，现代的设计手法，在“棱形”过桥上体现的这一点，棱形过桥的桥底是 GRG 发散性的冲孔发光图案，到桥的最低点，有机的图案点汇聚成无机的图案点，最终形成一个陕西皮影戏人物的剪影图案，使得整个商场的室内氛围别具一格。

Waved shape of the long gallery directly corresponds to each bridge, and all bridges are arranged at the lowest point of waves, which is a unique design. Based on this, undulating waves on lateral wall and streamline patterns have variations. "Prismatic" bridge of the long gallery is no doubt the "finishing touch" of the whole long gallery design. The bridge is shaped like a half "prism" lightly inlaid in the long gallery. The center of the bridge deflects to one side, adding vitality to the otherwise clumsy bridge. Circular inserted light makes the bridge lighter and with a modern sense. A mix of traditional culture and modern design method is reflected in the design of "prismatic" bridge, with GRG emanative and perforated glowing patterns at the bottom of the prismatic bridge, and organic pattern points converge to inorganic pattern points at the lowest point of the bridge, thus finally forming a sketch pattern of a Shaanxi shadow play figure and making the interior atmosphere of the whole store distinctive.

20 室内步行街连桥底天花平面图
21 室内步行街连桥底侧裙立面图
22 室内步行街连桥

21

VERO MODA

24

23 室内步行街过桥底特写
24 室内步行街过桥底窝蛋造型

25

LANDSCAPE OF PLAZA
广场景观

景观设计通过挖掘当地历史文化，结合建筑立面造型而得出理念——穿越时空的繁华，以一只唐服上的蝴蝶为线索，从唐朝穿越时空来到现代，栖息于这片土地，见证了从古到今这片土地的繁华。设计中结合建筑立面的形态和蝴蝶翅膀的纹路，运用于主入口铺装、绿化、小品、城市家具等。

Based on exploration on local history and culture and by taking into account architectural elevation shape, the plaza landscape design concept takes a butterfly on Tang Dynasty's costume as clue, which comes to modern time from the Tang Dynasty, inhabits on the land and witnesses the prosperity of the land since ancient times. Architectural elevation shape and butterfly wing pattern are adopted in main entrance pavement, greening, featured landscape and urban furniture.

25 广场景观设计手稿
26 广场主雕塑
27 广场景观
28 绿化花坛

万达广场 倾城绽放
万达广场 倾城绽放
万达广场 倾城绽放
11月22日
盛大开业
入口
永辉超市
11月22日
入口
H&M

30a

30b

30c

29 广场座椅
30 广场主题雕塑

31

31 广场全景夜景
32 百货立面夜景

NIGHTSCAPE OF PLAZA
广场夜景

建筑夜景照明设计之初，在成本分配上下功夫，确定了重点表达部位次序：大商业主入口，购物中心主立面（西），蜡烛顶部，蜡烛主立面（西）。设计过程中严格执行限额设计，确保了最终没有突破集团建造标准。划分夜景照明渲染重点，也是照明效果艺术性的需要。作为大型商业建筑群，需要考虑城市尺度、街区尺度以及近人尺度的观看效果。塔楼屋顶的天际线的重点刻画，购物中心立面飞天的重彩描绘，以及购物中心入口的LED视频点幕，不论从艺术性还是观赏距离把握上，都体现设计者深厚的功底。

At the beginning of nightscape lighting design, parts to be expressed as priority are determined in the following sequence according to cost allocation: main entrance of large commercial center, main elevation (west) of the shopping center, candle top and main elevation of candle (west). Quota design is strictly followed during the plaza nightscape design so as to meet the Group's construction standard. Determining nightscape lighting priority is necessary as it can demonstrate artistic feature of lighting effect. Nightscape design for the large commercial building complex is made by considering viewing effects in terms of urban scale, block scale and intimate scale. Design of skyline at the tower building roof as a priority, strong coloring of the shopping center elevation depicting the image of "flying to the sky" and LED video point curtain at the shopping center entrance all reflect competence of designer in both artistic design and viewing distance determination.

32

33

万达影城
WANDA CINEMA
毛衣
RMB199
H&M GRAND OPENING IN NOVEMBER!
11月，盛大开业!

万达广场
WANDA PLAZA

35

OUTDOOR PEDESTRIAN STREET
室外步行街

室外步行街的立面设计上，重点刻画细部处理，简洁的现代造型与抽象的当地文化符号相结合；不同效果的门头处理、仿古砖主墙面和铝板外墙墙面，交错辉映，"一店一色"，有机结合，层次丰富，以细腻的手法表现现代美学，唤起人们对当地鼓楼文化广场的记忆，体现了现代与历史的辉映。

Outdoor Commercial Street elevation design focuses on detail treatment, featured by a combination of concise modern shape and abstract local cultural symbols; gates are treated by different ways, main walls are made of archaized bricks and exterior wall with aluminum plates, enhancing each other's beauty, "One store, one style", organically combined and rich in layers, adopting fine method to demonstrate modern aesthetics, arousing people's memory on local drum tower cultural plaza and demonstrating a blending of modern and history.

37

35 室外步行街门头
36 室外步行街转角
37 室外步行街立面

BENGBU WANDA PLAZA
蚌埠万达广场

开业时间 2013 / 11 / 29
建设地点 安徽 / 蚌埠
占地面积 18.31 公顷
建筑面积 95.06 万平方米

OPENED ON NOVEMBER 29 / 2013
LOCATION BENGBU / ANHUI PROVINCE
LAND AREA 18.31 hectares
FLOOR AREA 950,600 m²

OVERVIEW OF PLAZA
广场概述

蚌埠万达广场项目总占地18.31万平方米，总建筑面积95.06万平方米，集大型商业购物中心、商业步行街、精装SOHO、五星级酒店、甲级写字楼和高档住宅为一体，是引领皖北商业潮流的现代化城市综合体。

Bengbu Wanda Plaza project covers a total land area of 183,100 m² and gross floor area of 950,600 m². It integrates such functions as large shopping center, pedestrian street, finely decorated SOHO, five-star hotel, Class A office building and high-end residence, and represents a leading modern urban complex in north Anhui Province.

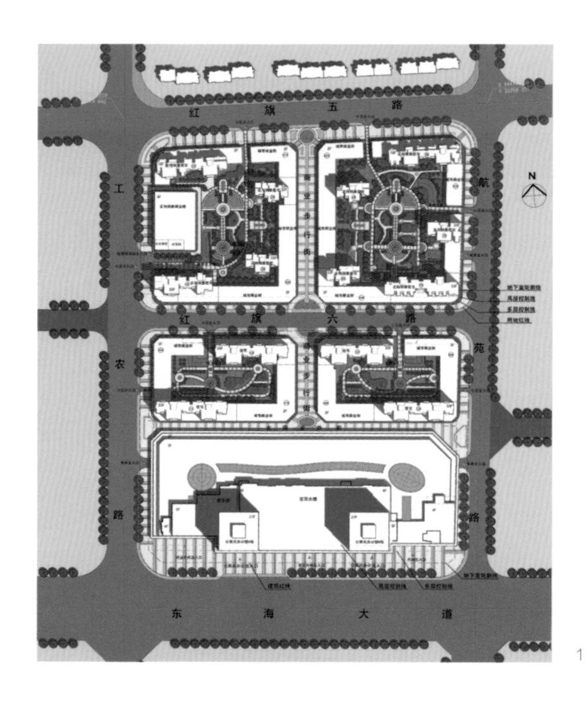

1

2

1 广场总平面图
2 广场鸟瞰图

万达广场
中心金铺 抢到赚到
万达广场
万达影城 IMAX
SEPHORA
盛大开业
Now open!

4

3 广场外立面
4 广场外立面正视图

FACADE OF PLAZA
广场外装

奔流不息的淮河文化，既为自然环境书写了新的奇观，同时也孕育了劳动者坚忍不拔的优良品质。跌宕起伏的横向条纹肌理为整个建筑演奏着变幻的韵律，使人仿佛身临其境，倾听着淮河水在蚌埠大地上洗刷着时间的印记。立面设计上的最大特点是横向带状肌理，通过层层出挑，曲折变化，赋予建筑跌宕起伏、充满活力的韵律。

Ceaseless Huaihe River culture not only writes new wonder of natural environment, but also breeds fortitudinous people. Fluctuating transverse stripe texture brings changing rhythm to the whole building, making one feel as if being on the scene, listening to the Huaihe River scouring imprint of time in Bengbu. The biggest feature of elevation design is transverse stripe texture, with layer-by-layer overhang, in twists and turns, endowing the building with fluctuating and vigorous rhythm.

入口的玻璃飘板，使得建筑立面虚实结合，灵动飘逸。穿孔板的运用，仿佛水面上闪烁的波光，赋予建筑十足的灵气。

Floating glass plates at the entrance reflects a combination of virtuality and reality on building elevation, flexible and elegant. Perforated plates, looking like glistening light of waves on water surface, endow the building with full anima.

5 广场一号入口
6 广场一号入口门头
7 广场二号入口

5

6

麦当劳

SEPHORA
盛大开业
Now open !
护肤 彩妆 香氛
打开美力
Beauty Power
TURN IT ON

8 广场一号入口

INTERIOR OF PLAZA
广场内装

本案结合地方文化及淮河文化，以水波涟漪为设计主题，以黑白灰为主基调，运用简洁、明快的设计元素，营造现代、高雅、时尚风格的综合体购物中心与城市区域中心。

门厅入口设计采用一体化设计，采用了透光玻璃和黑镜钢收边条。顶面流畅的线条，配合暖色灯带，使得墙顶造型贯通。

The plaza interior design is themed by water wave and ripple which demonstrate local culture and Huaihe River culture. Black and white gray is taken as main tone. Concise and vivid design elements are adopted to jointly create a modern, elegant and fashionable shopping center complex and urban regional center.

The lobby entrance adopts integrated design, with transmitting glasses and black mirror steel edging. Smooth top surface lines combined with light band in warm color form a cut-through wall top shape.

9b

9a

9 广场入口天花

10

11

12

圆中庭以简洁、明快的设计手法，庄重、大气的设计语言，强调商场公共空间作为一个良好的平台，充分体现商铺在商业空间主角的设计理念，营造高档、舒适的商业氛围。

The circular atrium design adopts a concise & vivid design method and solemn & grand design language, stressing public space as a good platform, fully reflecting the design concept of stores as leading player in commercial space and creating a top grade and comfortable commercial atmosphere.

10 圆中庭采光顶
11 圆中庭
12 圆中庭侧裙板

13

椭圆中庭设计采用以倒角弧线为设计元素，通过挑空侧裙板造型的变化、地面材质纹理的穿插过度、灯光的烘托等手法，营造大气、庄重、休闲的购物氛围。

Oval atrium adopts chamfer arc as design element, varying shapes of side plates on void, alternation and transition of floor material texture and light design jointly create grand, solemn and relaxed shopping atmosphere.

13 椭圆中庭
14 椭圆中庭采光顶

尚品宅配
BUFFET
万达广场
WANDA PLAZA

室内步行街侧裙板为白色GRG板，局部洗光，栏板使用安全玻璃配金属支架的设计，结构简洁、安全、稳固。

The indoor pedestrian street adopts white GRG side plates, with local light. Breast board made of safety glass and equipped with metal rack enables the structure simple, safe and stable.

15 室内步行街自动扶梯
16 连桥
17 室内步行街

15

16

17

LANDSCAPE OF PLAZA
广场景观

景观设计以“淮河生态”为主题贯穿始终，汲取建筑外立面设计的精髓，融入淮河地缘人文，用现代时尚的设计手法，展现场地“鸟寻花径知春到，鱼跃龙门带雨飞”的地域意境特征。

Themed by "Huaihe River biology", the plaza landscape design draws essence of building façade design, integrate local humanity into design and showcase a regional artistic conception of "birds seeking flower indicate that spring is coming, fishes leaping over the dragon gate in a rainy day" with modern design method.

19

18 广场景观绿化
19 特色花坛平面图

20a

20 广场主雕塑
21 景观小品
22 特色花坛

20b

21a

21b

21c

22

23

NIGHTSCAPE OF PLAZA
广场夜景

通过点、线、面不同的光源形式来塑造建筑在夜间的雕塑感和纯净感。在穿孔板背面设置点光源，特别是在近景观察，形成影影绰绰的光斑效果；在每条飘板下部吊顶根部设置LED泛光照明。在飘板之下形成时隐时现的金边光晕，托起整个飘板。面光效果由点光汇聚而成，最亮的地方希望出现在主入口和穿孔板密集区域，使之成为整个画面的视觉中心。

Different light sources (point, line and surface) are utilized to create a sculpted and pure building image at night. Point light sources are set at the back of perforated plates so as to generate a shadowy light spot effect; LED floodlighting at the root of the suspended ceiling below each floating plate form golden edge halo below the plate and hold up the whole plate. Ceiling spotlight effect is realized through gathering of point lights, and it is better that the main entrance and area where perforated plates are densely arranged are brightest so as to become visual center of the whole picture.

23 广场外立面夜景

24 主入口夜景

万达广场
WANDA PLAZA
麦当劳
watsons
屈臣氏
hotwind
珠城首席

26

27

28

25 门头夜景特写
26 广场水景夜景
27 广场夜景
28 二号入口夜景

OUTDOOR PEDESTRIAN STREET
室外步行街

室外步行街的设计延续“淮河生态”的设计主题，将各种元素合理组合、布置得当。所有元素的布置，寻求空间的序列和秩序，在有序与变化中寻求平衡。

The outdoor pedestrian street design also reflects the design theme of "Huaihe River Biology", featured by rational combination and allocation of various elements. Spatial sequence and order are taken into consideration in elements allocation to strike a balance between orderliness and variation.

29 室外步行街入口
30 室外步行街立面

30

XUZHOU YUNLONG WANDA PLAZA
徐州云龙万达广场

开业时间 2013 / 12 / 06
建设地点 江苏 / 徐州
占地面积 12.84 公顷
建筑面积 47.11 万平方米

OPENED ON DECEMBER 6 / 2013
LOCATION XUZHOU / JIANGSU PROVINCE
LAND AREA 12.84 HECTARES
FLOOR AREA 471,100 m^2

OVERVIEW OF PLAZA
广场概述

徐州云龙万达广场是徐州市的“三重一大”重点工程，云龙核心区的龙头项目。项目地处云龙区和平路与庆丰路交会处，规划总建筑面积47.11万平方米。整个项目由大型购物中心、城市商业街、商务写字楼、豪华国际影城和高档住宅等多种业态组成，是融购物、餐饮、文化、娱乐、商务、休闲及居住等多种功能于一体的大型城市综合体。

Xuzhou Yunlong Wanda Plaza is one of key projects of Xuzhou under the "three important and one large" (major problem decision making, important cadre appointment, key project investment decision making and large fund utilization) plan, and a leading project of Yunlong Core Area. Located at the junction of Heping Road and Qingfeng Road in Yunlong District, the project has a gross floor area of 471,100 m^2, and is composed of various business types, including large shopping center, urban commercial street, business office building, luxury international film cinema and high-end residence. The project represents a large urban complex integrating such functions as shopping, F&B, culture, entertainment, commerce, leisure and residence.

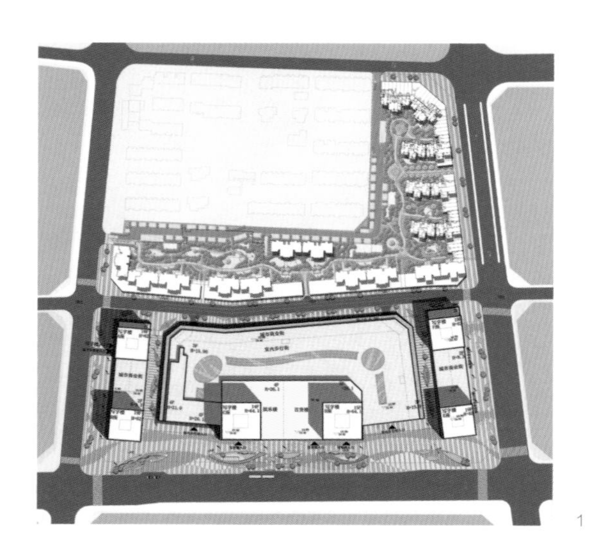

1

1 广场总平面图
2 广场鸟瞰图

2

FACADE OF PLAZA
广场外装

徐州云龙万达广场大商业建筑部分通过双层网格板与金银飘带的交织组合，抽象地表现了云龙的设计理念。塔楼部分为了呼应大商业建筑的流动性采用弧线造型，着重表现龙在云中穿梭的姿态。大商业建筑整体完美地表现出云龙区固有的文化特色；同时辅以万达蓝色的门头设计，成为整个项目的点睛之笔。

Inter-weaved with double-layer grid plates and gold & silver ribbons of the large commercial building abstractly reflect the design concept of dragon passing through clouds. Arc-shaped tower building corresponds to mobility of large commercial building, highlighting posture of dragon passing through clouds. The large commercial building as a whole perfectly demonstrates cultural features of Yunlong District, supported by blue gate design which is unique for Wanda, becoming the finishing touch of the project.

3 广场外立面

4

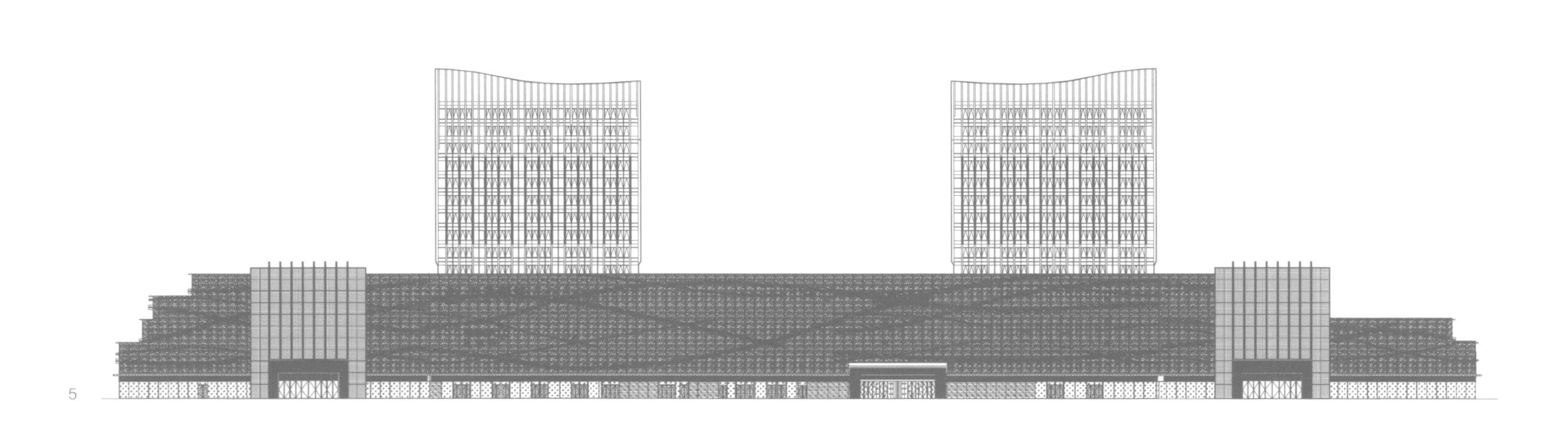

5

6

大商业建筑门头采用万达蓝固有色，色彩上突出建筑的归属性。在构造处理上，外表皮采用蓝色低透玻璃，中间层为穿孔铝板，内层藏灯。门头上方八条竖向弧线飘带抽象出龙在天空中的理念，与裙房的飘带造型形成了很好的呼应。

大商业建筑外立面的大、小飘带为流线式造型的铝板，金银交织，如包裹礼物的彩带使得立面效果独一无二；起伏的塔楼顶部轮廓呼应整体曲线外形；主立面为双层镂空铝板的幕墙，宏伟的建筑不失体量感，并体现了细腻丰富的视觉效果。

Gate of the large commercial center is in blue, an iconic color of Wanda Group, thus stressing belongingness of the building. The building structure consists of blue low-permeability glasses on the external surface, perforated aluminum plates on the middle layer and inserted light in the interior layer. Eight vertical arc-shaped ribbons above the gate showcase the concept of dragon in the sky, and well correspond to the ribbon shape of the podium.

Facades of the large commercial center are designed with large and small ribbons made of streamlined aluminum plates, with gold and silver ribbons interweaved, thus generating a distinctive facade effect which looks like color ribbons for packing gifts. Fluctuating contour of the top of the tower building echoes with the overall curve shape. Curtain wall on the main facade is made of double-layer hollow out aluminum plates, giving the magnificent building a sense of size and reflecting a fine and rich visual effect.

7a

7b

4 广场二号入口
5 广场立面图
6 百货外立面
7 百货外立面特写

8

9

INTERIOR OF PLAZA
广场内装

室内设计方案以云朵为主基调，通过简约、素雅的手法，以行云流水的形态进行设计，其中最为精彩的是天花设计，曲径通畅的造型灯带首尾衔接，连贯一体，在购物中心室内空间形成一气呵成的意境，也增加了空间的趣味性。

主入口、门厅作为室内步行街的起点，也是内装元素开始的地方。天花造型元素也从此区域开始切入，形成很强的方向感，达到吸引顾客进入室内步行街的目的。

By taking cloud as main tone, the interior design adopts simple and elegant method to realize an image of floating clouds and flowing water. The most wonderful design is for suspended ceiling, unobstructed modeling light bands are connected in a head-to-tail way, thus creating a coherent atmosphere in the shopping center and making the space more interesting.

The interior finishing elements are adopted starting from the main entrance and lobby, the starting point of the indoor pedestrian street. Suspended ceiling modeling elements are adopted starting from this area so as to generate a strong sense of direction and realize the goal of attracting customers.

10

8 入口门厅
9 入口门厅特写
10 入口天花

椭圆形中庭侧裙板和天花采用相同的线条元素，由强烈的韵律感灯带造型与侧裙板线条进行呼应，配合纯白色的灯光效果，从而营造出简约宁静、飘逸律动的购物空间氛围。

Lateral wall and suspended ceiling of the oval atrium adopt the same stroke elements, utilizing highly rhythmic light band shape to correspond to strokes of side skirt plates and designing with pure white light effect, thus creating simple, tranquil, elegant and rhythmic shopping space.

11

12

11 椭圆中庭设计手稿
12 椭圆中庭采光顶
13 椭圆中庭

mothercare
mothercare
SEPHORA
SEPHORA
GUESS
Calvin Klein Jeans
UNI
QLO

13

輝煌万
徐州

14 圆中庭采光顶

15

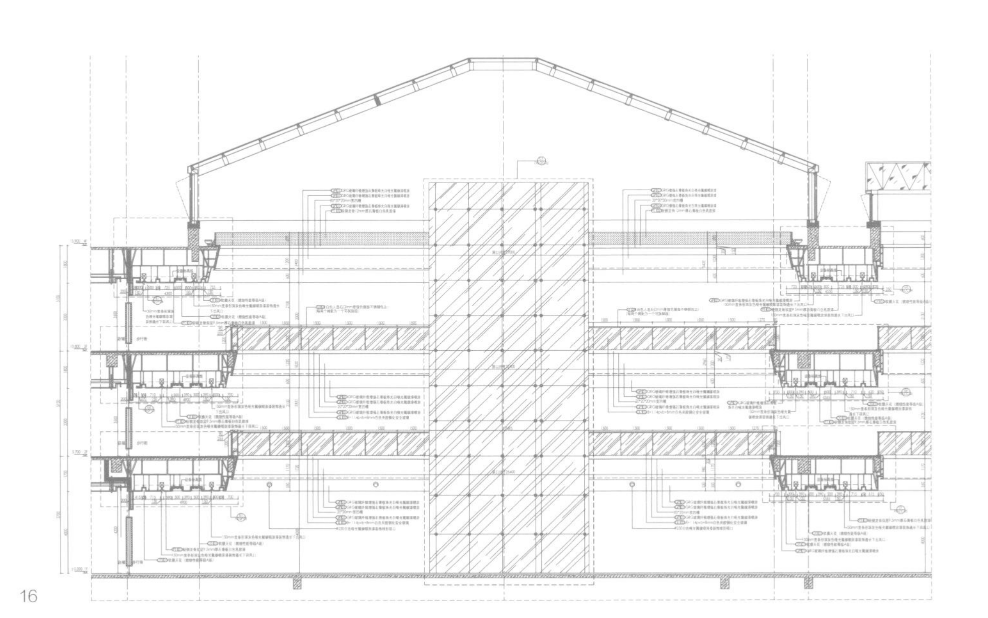

16

15 圆中庭
16 圆中庭剖面图
17 圆中庭雕饰
18 圆中庭仰视

17

18

圆中庭的设计充分运用了建筑本身的圆形结构，同时在原结构基础上增加了挑台，犹如形态各异的云朵，强化了空间的呼应关系，也形成了购物中心内装的亮点。

Circular structure of the building is fully utilized in circular atrium design. Balcony is added on the existing structural foundation, which seems like clouds of different shapes, thus enhancing responding relationship of the space and highlighting the interior design of the shopping center.

19

在内装设计中，通过调整连桥、自动扶梯的空间布置，与内装主题元素紧密结合，实现连绵不断、上下起伏，精致中又富于变化的空间效果。在空间上，直街上下空间连续呼应，达到一气呵成的意境。

In the interior design, overhead bridge and escalator layout is adjusted according to the interior design theme elements, thus generating a ceaseless, fluctuating, delicate and changing spatial effect. The upper and lower parts of the straight gallery correspond to each other continuously, thus generating a coherent effect.

20

19 连桥
20 室内步行街设计手稿
21 室内步行街
22 自动扶梯

21

22

23

LANDSCAPE OF PLAZA
广场景观

依据汉代乐器编钟的造型，提炼出其抽象化、艺术化的雕塑小品造型，上下起伏的发光体就像跳动的音符一样，在五彩缤纷的光线下给人们塑造一种浪漫而又温馨的轻松购物空间。

Abstract and artistic featured sculpture shape is extracted from the shape of chime, a musical instrument of the Han Dynasty. Fluctuating illuminant, looking like jumping notes, emits colorful lights and offers a romantic, cozy and relaxed shopping space.

23 广场主题雕塑
24 广场主雕塑

24

25

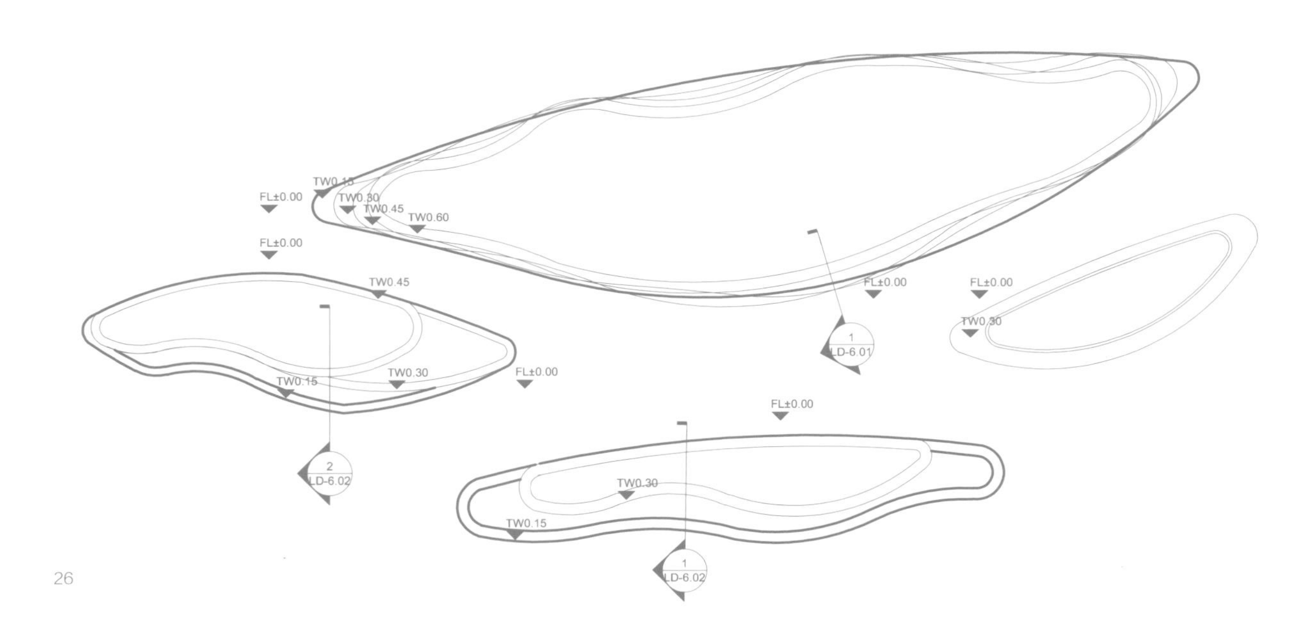
FL±0.00
TW0.15
TW0.30
TW0.45
TW0.60
FL±0.00
TW0.45
TW0.15
TW0.30
FL±0.00
FL±0.00
FL±0.00
TW0.30
1
LD-6.01
FL±0.00
2
LD-6.02
TW0.30
TW0.15
1
LD-6.02

26

27a

27b

25 广场景观绿化
26 特色花坛组合布局图
27 特色花坛

万达广场
WANDA PLAZ
STARBUCKS COFFEE

28 广场夜景

29

30

29 广场二号入口夜景
30 广场二号入口灯光幕墙
31 广场全景夜景

31

NIGHTSCAPE OF PLAZA
广场夜景

徐州云龙万达广场作为徐州市云龙核心区的龙头项目，也将夜景照明打造成区域的地标性灯光景观。幕墙为夜景照明提供了很好的预留照明空间及平台，以云、龙、水为设计元素。借助绚丽的灯光动画表现形式，重在体现建筑本身的特质及纹理走向。灯光设计旨在烘托商业氛围，提升建筑空间品质；灯光设计主题为“龙鳞隐波，时尚彭城”。

建筑的主要焦点为大商业裙楼立面的金属网板及一条条柔美的线条结构。为了体现建筑的立面特色，夜景照明设计在不锈钢网板背面安装LED线性洗墙灯打亮背板，通过间接光反射出来，很好地隐藏了灯具及管线；立面铝板曲线造型端面的凹槽中隐藏安装LED线性洗墙灯，层次分明。动画式的表现效果犹如一条条蛟龙在云中游动，给人以强烈的视觉冲击力及热闹的商业氛围。

As a leading project of Yunlong Core Area in Xuzhou, nightscape lighting design of Xuzhou Yunlong Wanda Plaza is a landmark light landscape in the region. Curtain wall offers a good lighting space and platform for nightscape lighting. By adopting cloud, dragon and water as design elements, the nightscape design focuses on expressing distinctive features of the building and its texture direction by means of gorgeous light animation presentation. Themed by "dragon skin hidden in waves, fashionable Xuzhou", light design is aimed to set off commercial atmosphere and elevate spatial quality of the building.

Main focuses of the building are metallic grid plates and soft strokes on podium building elevations of the large commercial center. In order to reflect building elevation features, LED linear flood lights are installed at back side of stainless steel grid plates, concealing well lighting fixtures and pipelines by reflecting light indirectly; LED linear wash wall lamps are concealed in groove at the end face of curved aluminum plates on elevation, in a structured way. Through animation manifestation, an image of dragons swimming in clouds is created, which creates a strong visual impact and busy commercial atmosphere.

OUTDOOR PEDESTRIAN STREET
室外步行街

室外步行街外装设计采用与大商业建筑群契合的竖向构件与背侧玻璃交错的形式，从而抽象表现龙在云间的理念，并运用大量大尺度飘板活跃天际线。东西两条室外步行街成为文化展示的重要载体，室外步行街的设计理念来源于整个建筑外立面的设计风格，室外步行街区域作为休闲娱乐的小尺度空间，从集中展示、立体重现、恢复与弘扬等三个层面来传承汉文化，使人们在感受浓郁的汉代文化与历史信息的同时，又可以感受现代时尚的休闲氛围。

主题雕塑、户外家具将成为万达广场汉文化集中显现区，采用现代材料与传统元素相结合的方式，创建一个适合步行、充满生机的城市街区。

In façade design of the outdoor pedestrian street, vertical members and glasses on back side are staggered to echo with the large commercial building complex and manifest the concept of dragon in cloud in an abstract way. Besides, a plenty of large-sized floating plates are arranged to make skyline dynamic. As important carriers of cultural presentation, the east and west outdoor pedestrian streets are designed on the basis of the design style of the whole building facade. The outdoor pedestrian street area is designed as a small leisure space to carry forward the Sinic culture from three perspectives, namely, concentrated demonstration, three-dimensional representation and resumption & promotion, thus creating a modern and fashionable leisure atmosphere where people can understand Han culture and historical information.

The Sinic culture will be demonstrated through themed sculpture and outdoor furniture which are designed by combining modern material and traditional elements, thus creating an urban block which is full of vitality and suitable for walking.

32 室外商业街入口

JOYGOGO
DIADROA
MONT BLAC
DORME
LENOD
OIMD
OIMD
OIMD
F-SHOP
SUGARS
TOUCH

33

33 室外步行街入口商铺
34 室外步行街夜景
35 室外步行街外立面
36 商铺门店夜景

STYLE
STYLE
VICTIA
VICTIA
FREMKE
FREMKE
GERIES

34

35

Toamel
Tomkjoly
Tomkjoly
Volare

DANDONG WANDA PLAZA
丹东万达广场

开业时间 2013 / 12 / 20
建设地点 辽宁 / 丹东
占地面积 26.07 公顷
建筑面积 111.07 万平方米

OPENED ON DECEMBER 20 / 2013
LOCATION DANDONG / LIAONING PROVINCE
LAND AREA 26.07 HECTARES
FLOOR AREA 1,110,700 m^2

OVERVIEW OF PLAZA
广场概述

丹东万达广场项目位于丹东市振兴区，由桃源街、春一路、人民街、花园路4条市政道路的围合地块内，是集五星酒店、写字楼、大型购物中心、商业街及高级住宅区等多种业态于一体的高端城市综合体，总建筑面积111.07万平方米。商业部分建筑面积28.0万平方米，其中地上面积20.0万平方米，地下面积8.0万平方米。

Situated in Zhenxing District of Dandong, Dandong Wanda Plaza is enclosed by four municipal roads (Taoyuan Steet, Chunyi Road, Renmin Street and Huayuan Road). It represents a high-end urban complex integrating such functions as five-star hotel, office building, large shopping center, commercial street and exclusive residential district. The project, covering a gross floor area of 111.07 m^2. The floor area of the commercial area is 280,000 m^2, including 200,000 m^2 for aboveground area and 80,000 m^2 for underground area.

1

1 广场总平面图
2 广场鸟瞰图
3 广场设计手稿

2

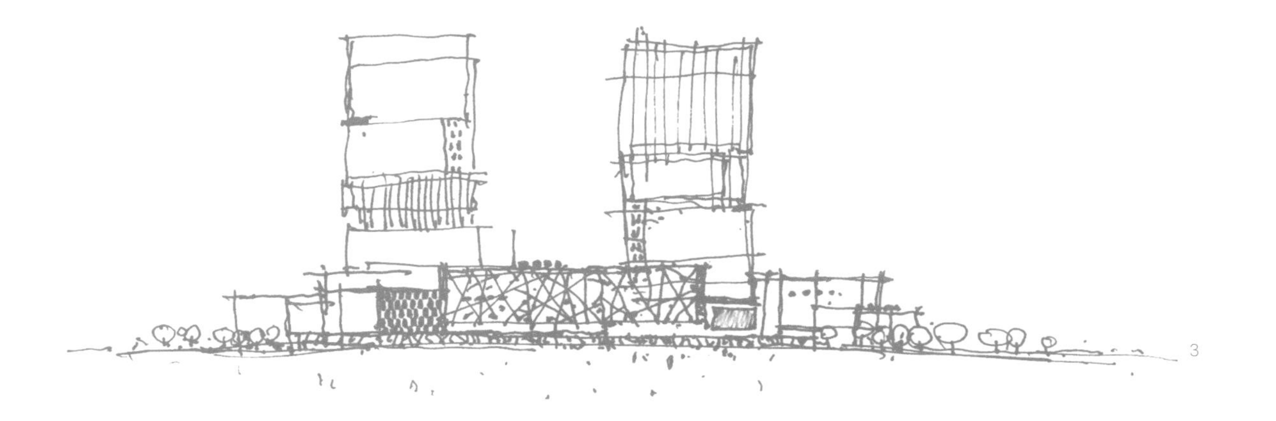

3

4a

4 广场外立面

FACADE OF PLAZA
广场外装

本项目在设计构思和建筑造型上，大胆引入了“礼盒”这一基本设计理念，从大商业裙房到酒店、办公塔楼，都采用了通过不同尺度、比例的方形体块堆叠方式作为本项目建筑立面设计中最基本的“母题”，在视觉效果上营造高度统一的设计风格，强化整体建筑群体的地标性。为体现建筑整体形象并衬托和突出立面细节，在建筑外立面的用色上，采取以白色、珠光白和浅灰为主要基色（背景色），商业裙房各体块的表皮细节采用强对比的明快色彩和图案的处理手法，不仅使得建筑整体形象在日间更为突出，也为夜间照明提供良好的界面和表现基底，同时，亦有考虑到当地朝鲜族文化，特别是传统民族服饰中，特有崇尚白色并配以高纯度鲜艳色彩映衬地域性审美需求。

The project design and building shape originate from the basic concept "gift box". Elevation design for the podium of the large commercial center, hotel and office tower building adopts the "basic theme" of stacked square blocks of different sizes and scales, creating a highly unified design style in terms of visual effect and highlighting the orientation of the whole building complex as a landmark. In order to reflect the overall building image and stress elevation details, the building façades mainly adopt white, pearl white and light gray as primary colors (background colors), surfaces of different commercial podium blocks adopt bright colors and patterns with strong comparison effect, which not only highlights the overall building image at daytime, but also provide a good interface and manifestation basis for night lighting. Meanwhile, giving the culture of the Korean nationality, especially valuing of white color in traditional national costume, bright colors with a high purity are added to reflect regional aesthetic demand.

4b

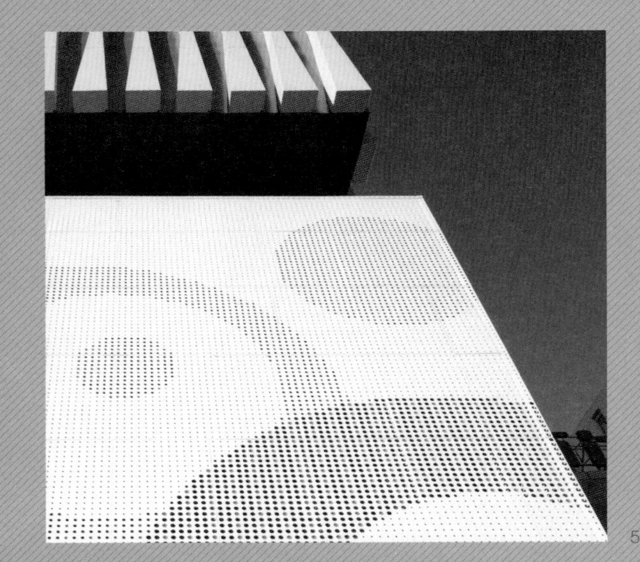
5a

5b

5c

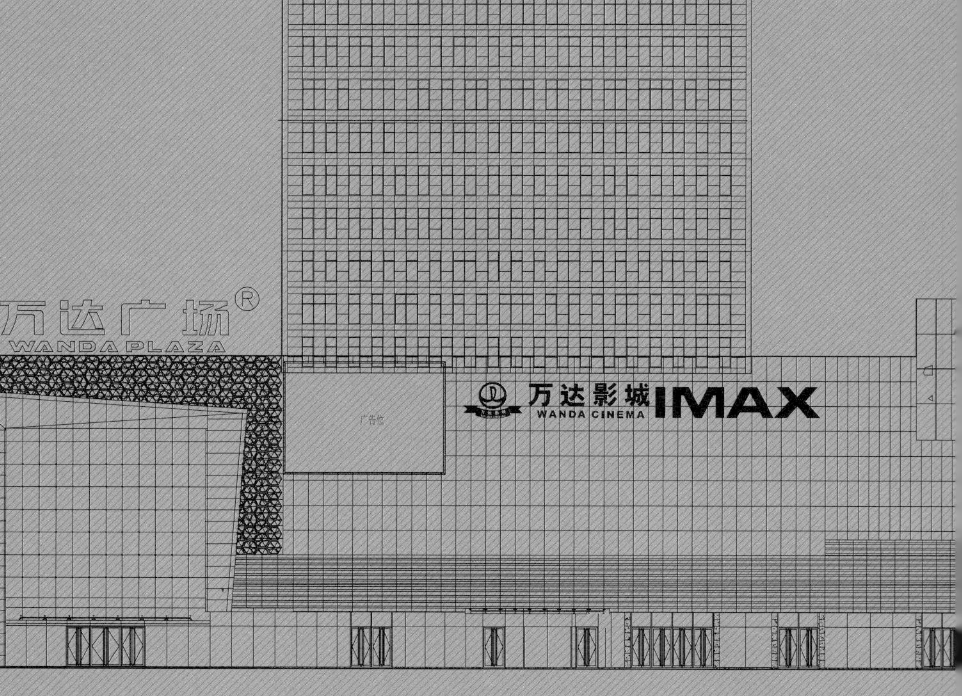

6

5 外立面局部特写
6 广场立面图

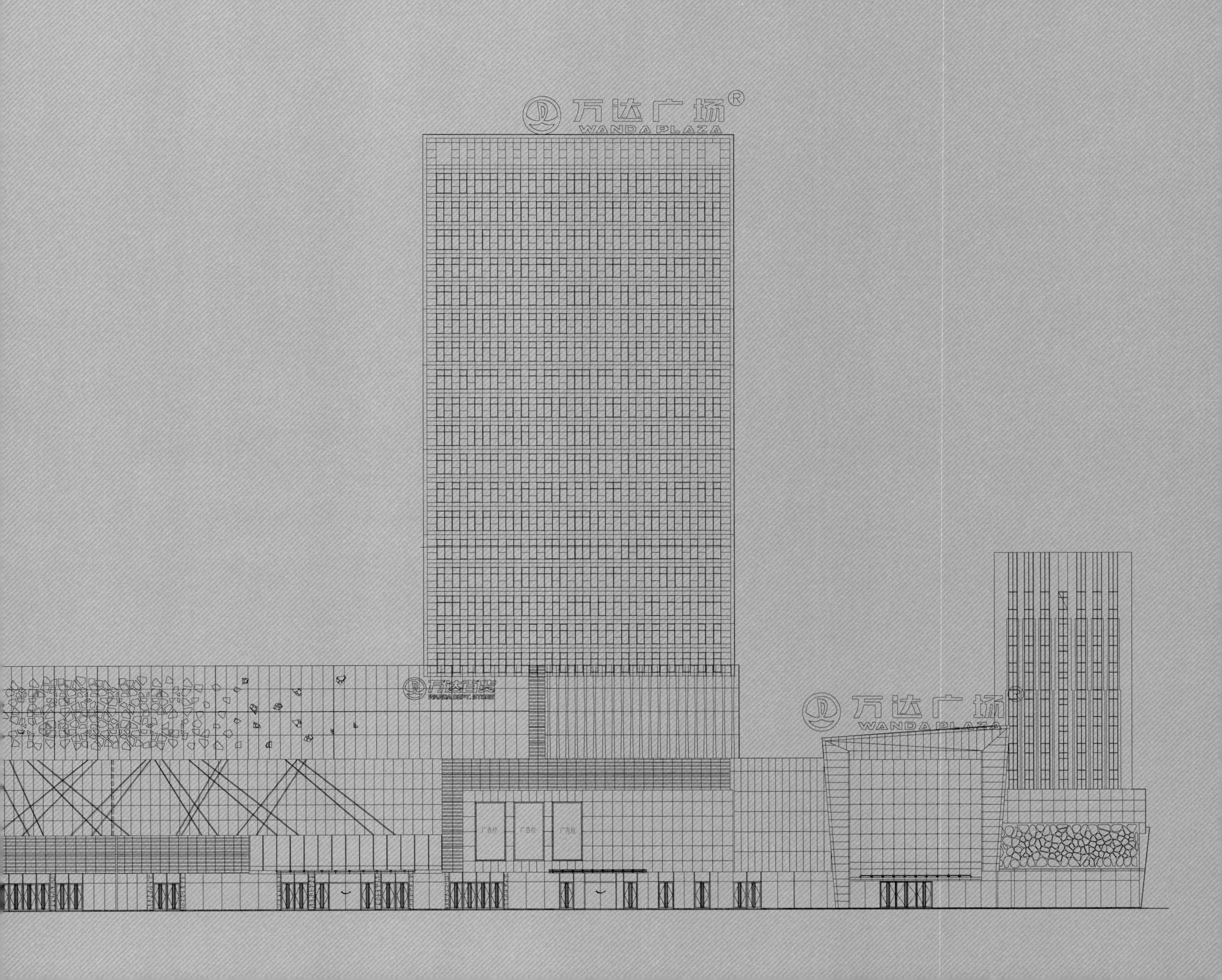
万达广场®
WANDA PLAZA
万达广场®
WANDA PLAZA

7

8

7 外立面局部特写
8 广场外立面
9 百货立面图

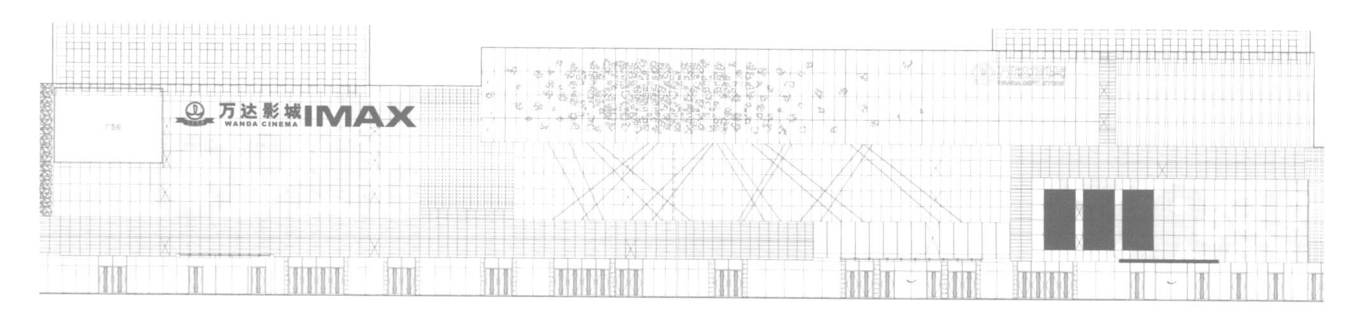

9

大商业的主入口的造型门头是整个建筑立面的设计重点，两个门头的造型左右对称，采用鲜亮的橘色铝板和“手工折纸”般异形斜面翻边的造型；门头内侧为3层通高的超白玻璃幕内衬硅钙板图案的双层幕墙，内层的硅钙板幕墙采用浮雕和编织的几何图案肌理，为夜景动画提供背景界面。

Shape design of gate at the main entrance of the large commercial center is the key of the overall building elevation design. Two gates are symmetrically arranged with bright orange aluminum plates and upturning irregular oblique plane shape which looks like "paper folding". The interior sides of the gates are arranged with 3-storey high double-layer curtain wall made of ultra-clear glass, lined with calcium silicate boards. The interior calcium silicate board curtain wall is designed with embossment and woven geometric patterns, thus offering background interface for nightscape animation.

10

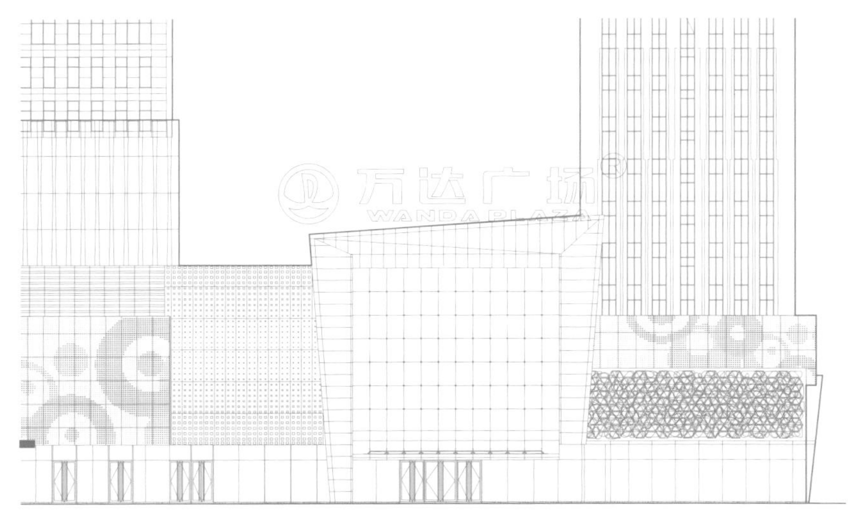

11

10 广场一号入口门头
11 主入口立面图
12 广场一号入口

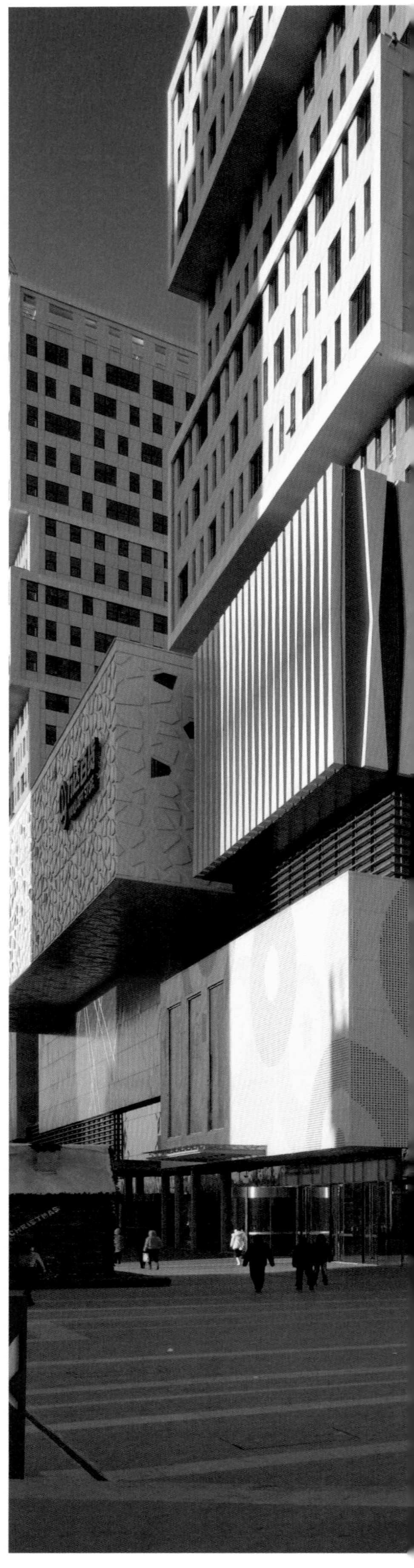

12

万达广场

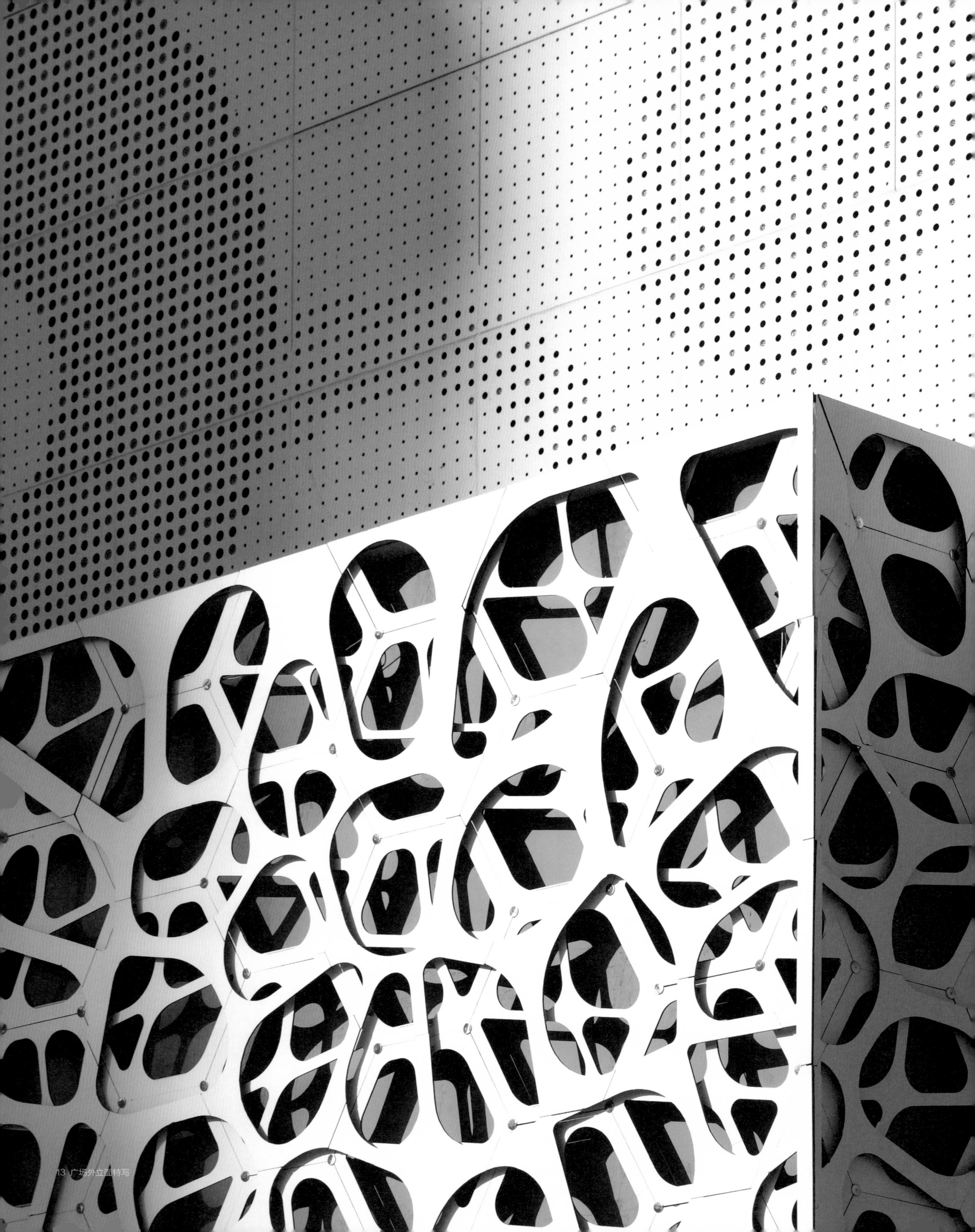
13 广场外立面特写

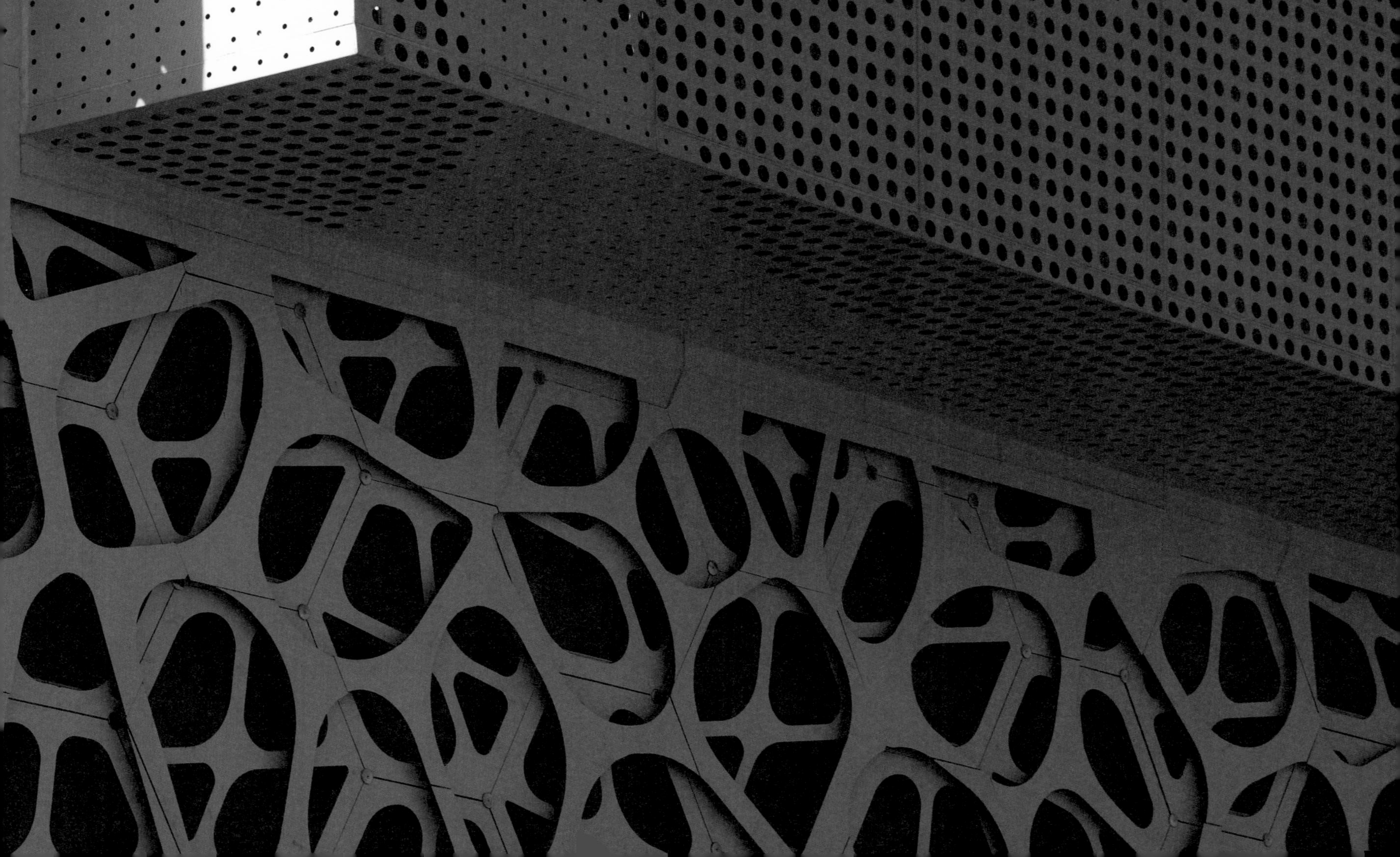

14

INTERIOR OF PLAZA
广场内装

内装设计上以丹东地域特色的鸭绿江、鸭绿江大桥、断桥为起点，引申出纵横交错的直线与弧线的交错变化，使得空间变化丰富、灵动，造型大胆又不失和谐之美，在空间色调上以简洁的白色为主基调，配合局部造型的色调变化，并以灯光的辅助勾勒出丹东万达室内步行街时尚、活泼、轻松的购物氛围。入口以光为线条，勾勒灵动的空间。光带由线及面的渐进，引人入胜，带入视觉的吸引与留恋。

The interior design of the Plaza takes the Yalu River, Yalu River Bridge and Yalu River Broken Bridge as the starting point, with alternative changes of crisscrossed straight lines and arc lines, forming a flexible space with variations, in a bold but harmonious shape. Concise white as the main tone, tonal variations on local shape, and light design jointly create a fashionable, lively and relaxed shopping atmosphere. The entrance outlines a flexible space with lights as strokes. Progressive design of light band from line to surface is attractive, with a good visual effect.

15a

15b

14 入口天花
15 主入口门厅

圆中庭金色的观光电梯，弧线的贴膜图案，犹如燃烧的火焰，交错的侧裙板金属贴片元素引自鸭绿江大桥的上部造型，加以提炼。观光电梯、侧裙板与地面交错的弧线拼花预示着此处将成为顾客汇集的交点。

Golden panoramic lift in the circular atrium is designed with arc-shaped filmed patterns, which looks like burning flame. Staggered metallic stickers on side plates are a refinement of the upper part shape of the Yalu River Bridge. Panoramic lift, side plates and crisscrossed arc floor patterns all indicate that this space will be a point where customers gather.

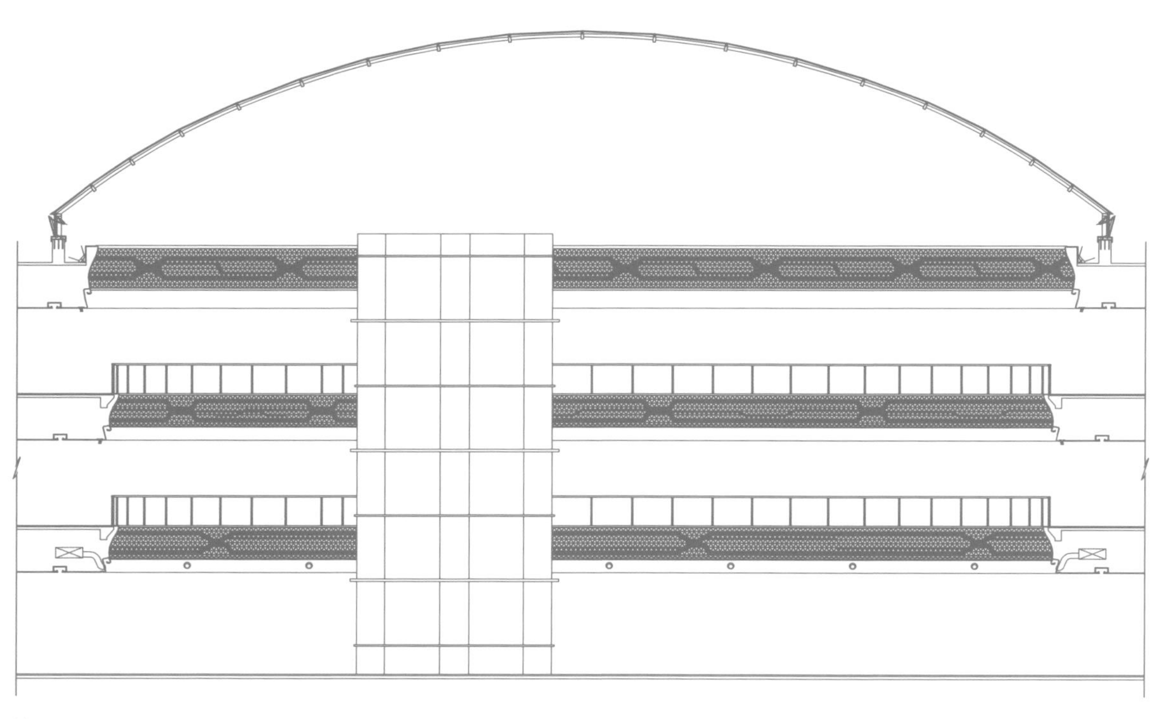

16

16 圆中庭剖面图
17 圆中庭

18

18 椭圆中庭采光顶
19 椭圆中庭

椭圆中庭以深蓝色的观光电梯配以银灰相间的贴膜图案，使得空间上仿佛趋于平静，侧裙板的金属贴片亦来自鸭绿江断桥的元素提炼，地面的弧形拼花，宛如静静流淌的河水，静候下一次绚丽的开始。

Dark blue panoramic lift and silver-and-gray film patterns in the oval atrium generate a quiet space; metallic stickers on side plates are extracted from element of Yalu River Broken Bridge. Arc-shaped floor patterns appear like silently flowing river waiting for another gorgeous moment.

20

21

22

20 室内步行街
21 连桥
22 自动扶梯
23 室内步行街侧裙板展开图
24 室内步行街采光顶

23

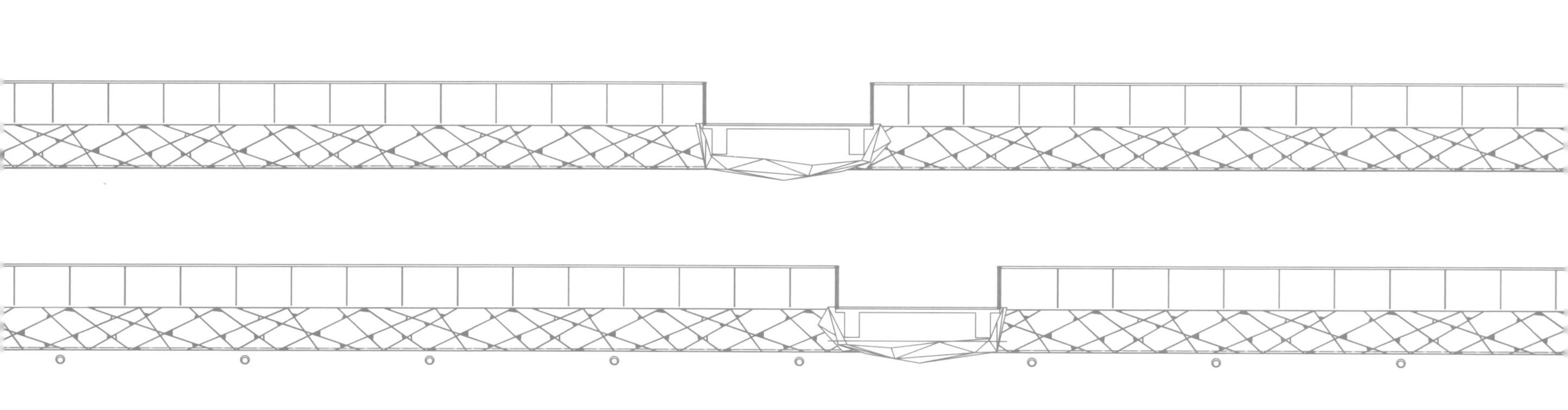

24

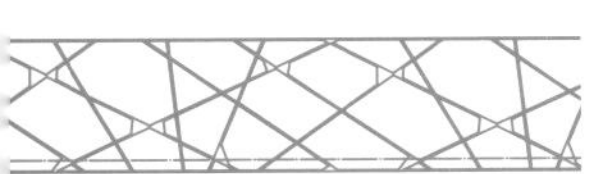

室内步行街空间上的大弧形弯，使得空间具有很大的延伸性，配合弧面的侧裙板及拐角的弧形处理，整体简洁、协调统一、大气。天桥采用空间上交错的光线，勾勒出一个个变化的三角面，与侧裙板的交错线条有着异曲同工的呼应。

Large arc curve makes the outdoor pedestrian space more stretchable. Side plates are designed with arc surfaces and corners are cambered, thus creating a concise, harmonious and grand image. The overpass outlines many changing triangular facets by virtue of interlaced lights, which corresponds to staggered strokes on the side plates.

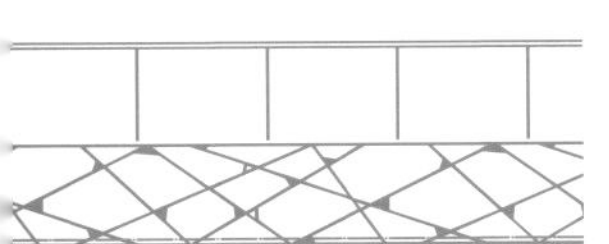

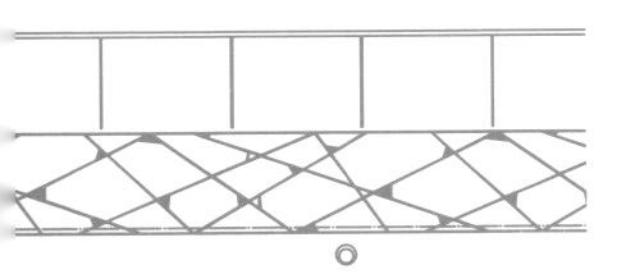

EACE
BIRD
太平鸟
1016
消火栓
FIRE HYDRANT
Shiny Crystal
它源于美丽的西班牙
精益求精的创作精神
让我们更爱自己

25 连桥

26

27

26 景观小品特写
27 圣诞主题景观小品
28 礼盒主题景观小品

LANDSCAPE OF PLAZA
广场景观

丹东万达广场整体景观设计风格紧密围绕建筑设计的“礼盒”主题；雕塑主题注重开业期的“圣诞”节日商业氛围营造。

Overall landscape design of Dandong Wanda Plaza is made to match by following the architectural design theme of “gift box”. The sculpture theme focuses on creation of commercial atmosphere as the Plaza opens during “Christmas” holiday.

28

29a

29b

主广场及商业街景观以丹东的地域风格“朝鲜”民俗主题为主。

The main plaza and commercial street landscape design mainly reflects folk theme of "Korean people" with regional characteristics of Dandong.

30a

29 朝鲜舞蹈主题雕塑
30 主题雕塑设计手稿

30b

NIGHTSCAPE OF PLAZA
广场夜景

将丹东的山水文化融入万达广场的夜景灯光中，采用国际现代手法，以山、水、光、影为组合的写意灯光，有着世外桃源般的意境，如同一幅韵味无穷的山水画，强调夜景灯光的丰富层次与艺术表现，力争给丹东的人们带来一种回归自然和超凡脱俗的感觉。

购物中心的主入口的灯具点阵是文化性表现的载体，突出入口，使之具有很好的开放性和吸引力。极具现代感的不规则外挑垭口飘檐内的轮廓灯带，通过折叠关系凸显形体的变化，创造出极具生命力的照明主题。

Landscape culture of Dandong is integrated into nightscape design of Wanda Plaza. Enjoyable light design combines mountain, water, light and shadow elements by means of international modern method, creating a land of idyllic beauty, which looks like a landscape painting of lasting appeal. Rich layers and artistic expression of nightscape light is a highlight so as to bring Dandong people a sense of back to nature and free from vulgarity.

Lamp matrix at the main entrance of the shopping center is a carrier of cultural expression. It highlights the entrance to make it more open and attractive. Contour light bands with a modern sense are set in overhanging eave of irregular pass, which highlights shape variations via folding relationship and creates a vigorous lighting theme.

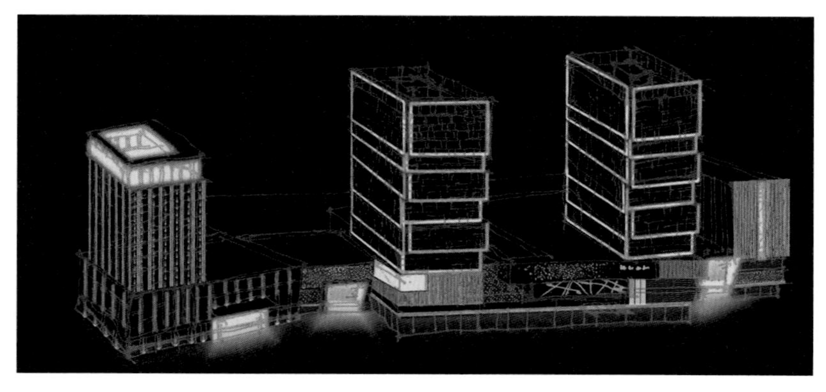

31

31 夜景灯光布局图
32 广场夜景

万达广场
WANDA PLAZA
万达百货
WANDA DEPT. STORE
万达广场
WANDA PLAZA

万达广场
broadcast

34

OUTDOOR PEDESTRIAN STREET
室外步行街

室外步行街延续堆叠错落的礼盒主题，采用珠光白、橘色的盒子构成，贯穿以通长的橘色铝板雨棚和暗红色横向铝格栅，与主立面整体效果相呼应的同时，雨棚、门套的预留灯槽和商铺间分隔墙的设置为夜景照明灯具、广告店招和美陈预留界面，以烘托室外步行街日间及夜间的商业气氛。

The commercial pedestrian street design follows the staggered “gift box” theme, adopting pearl white and orange boxes, with full-length orange aluminum plate canopy and dark red transverse aluminum grating, corresponding to the main elevation effect. Light trough reserved in the canopy and door pocket and partition walls among stores reserve interface for nightscape lighting fixture, ads shop signage and decoration, thus setting off the commercial atmosphere of the street during daytime and at night.

36a

36 室外步行街外立面
37 室外步行街入口

36b

37

36c

WANDA REALM
万达嘉华酒店 WANDA REALM

FACADE OF HOTEL
酒店外装

强化的竖向线条，有韵律的虚实对比，以及特别处理主楼屋顶造型，共同烘托出鲜明的酒店建筑性格。

Intensified vertical lines, metrical virtuality-reality contrast and special main building roof shape jointly create a distinctive hotel building.

39

38 酒店外立面
39 酒店入口

40

NIGHTSCAPE OF HOTEL
酒店夜景

设计把酒店分成上、中、下三个层次，顶部的城市天际线部分，中部的立面街道尺度部分和底部的人行尺度部分，立面照明设计的重点放在顶部上，使人们从远处就能领略它完整的丰姿，黄色的金顶饱满厚重，立面的竖向线条强调了建筑的挺拔和气质，底部的灯光构成则显示了酒店的尊贵。

The hotel nightscape design includes three parts: urban skyline at top, elevation street scale at central part and pedestrian scale at bottom. Elevation lighting is designed to make people enjoy its overall shape when standing afar, with rich and thick yellow golden top; vertical lines on elevations highlight the tall and straight building; bottom light demonstrates nobility of the hotel.

40 酒店入口夜景
41 酒店外立面夜景

WANDA REALM
万达嘉华酒店 WANDA REALM

WANDA REALM
万达广场
让城市更繁华
31188

万达广场
WANDA PLAZA
万达广场
万达广场
WANDA PLAZA

NANJING JIANGNING WANDA PLAZA
南京江宁万达广场

开业时间 2013 / 12 / 21
建设地点 江苏 / 南京
占地面积 10.11 公顷
建筑面积 53.30 万平方米

OPENED ON DECEMBER 21 / 2013
LOCATION NANJING / JIANGSU PROVINCE
LAND AREA 10.11HECTARES
FLOOR AREA 533,000 m^2

OVERVIEW OF PLAZA
广场概述

南京江宁万达广场项目位于南京市江宁区竹山路两侧，上元大街以南、规划道路及鼓山路以北、规划道路以东。南京江宁万达广场的总规划用地面积10.11万平方米，其中西侧建筑用地5.46万平方米，东侧建设用地4.65万平方米。规划总建筑面积约53.30万平方米。其中地上部分41.65万平方米，地下部分11.65万平方米。

Nanjing Jiangning Wanda Plaza is located on both sides of Zhushan Road in Jiangning District of Nanjing, to the south of Shangyuan Avenue, to the north of planned road and Gushan Road and to the east of the planned road. The project occupies a total land area of 101,100 m^2, including 54,600 m^2 west land and 46,500 m^2 east land area. The total floor area is 533,000 m^2, including 416,500 m^2 for aboveground part and 116,500 m^2 for underground part.

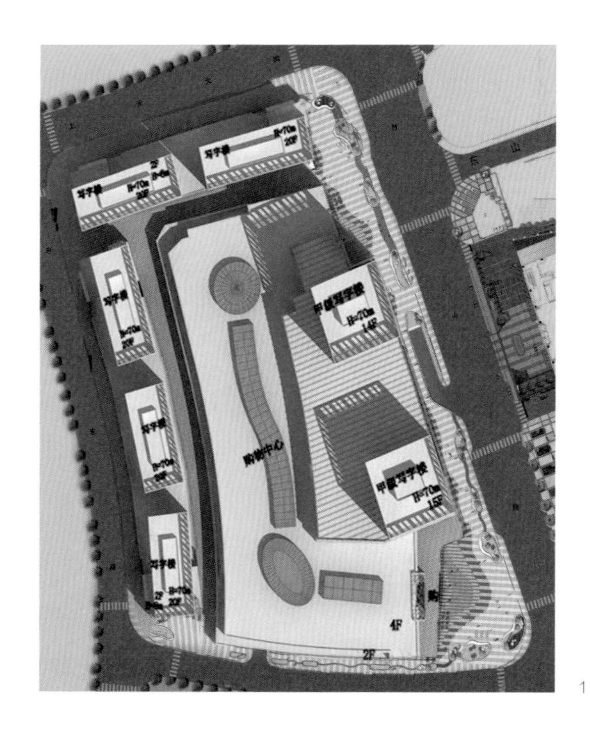
1

2

1 广场总平面图
2 广场鸟瞰图

3

FACADE OF PLAZA
广场外装

大商业采用曲线构图，交错重复，局部微差变化，获得了整体性强而又富有变化的效果，契合了南京滨水城市的自然特征，同时也符合万达简洁、完整、精致的设计定位。大商业入口以竖向波浪线条为主题，以通透的形象凸显入口的位置，与商业主体协调。

The large commercial center design adopts alternative and repeated curves, with local slight variations, thus realizing a highly integrated but variable effect. The design not only matches natural characteristics of Nanjing as a waterfront city, but also reflects Wanda's design orientation, i.e. simple, complete and delicate. The large commercial center entrance is themed by vertical wavy lines, highlighting the entrance position with a transparent image and coordinating with the main body of commercial center.

3 广场外立面
4 广场立面图
5 广场全景

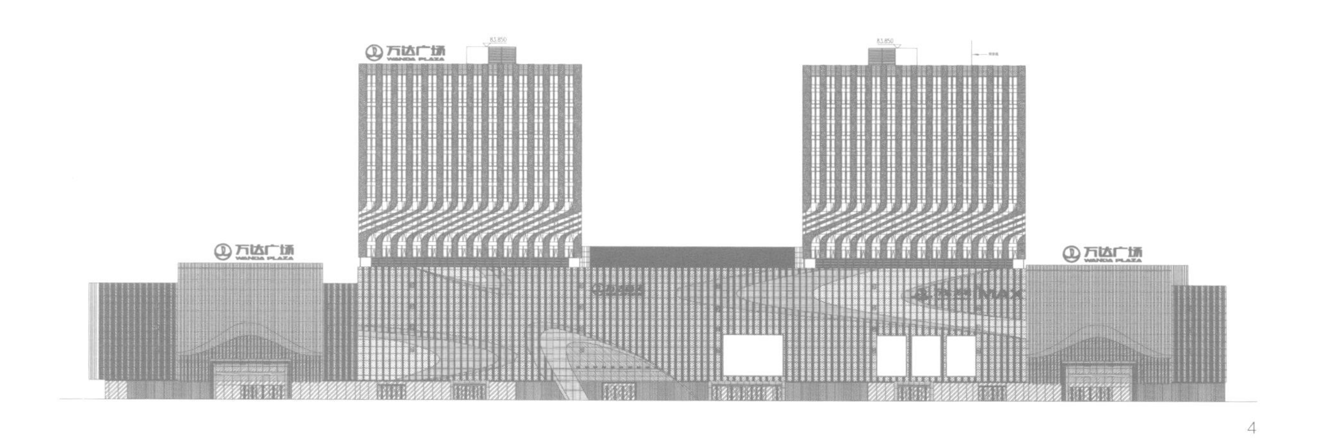

4

6 广场一号入口

永辉超市
1号门
永辉超市入口
2013.12.21
江宁万达
领越绽放
WANDA
PLAZA
GRAND
OPEN
3号门

7

INTERIOR OF PLAZA
广场内装

内装设计借助形体之间材质、肌理、色彩、灯光及疏密关系上的穿插变化，使空间简约而不单调，热闹而不喧杂。理性紧致的空间中加入空间构成，使之成为室内基本的符号元素；并结合本地特有的自然符号及城市文化，使空间整体氛围达到和谐统一，浑然天成的艺术效果。

In interior design of the Plaza, variations in materials, textures, colors, lights and density relationship create a simple but not monotonous space, which is lively but not noisy. Rational and compact space is added with spatial composition to serve as basic symbol elements in the interior, local natural symbols and urban culture are incorporated to create a harmonious and unified space and a totally natural artistic effect.

7 椭圆中庭
8 入口门庭
9 主入口天花

8

9

主入口设计考虑各部分的功能需求在狭长的横向通道天花设置一个LED天幕，由LED灯组成圆形图案，按照预计的频率开闭，最后形成了一个波光粼粼的水面效果，突出了以“水”为主题的购物环境。小厅的顶面为了增加引导性，设置了圆形的LED灯盒，内置钻石型的LED灯珠通过镜面的折射使圆形的灯盒更像夜晚的星空。

Based on functional demands of different parts, the LED backdrop is set on suspended ceiling of long and narrow transverse corridor in the main entrance. These LED lights form a circular pattern and are turned on and off at a planned frequency, realizing a glittering water surface effect and highlighting "water" themed shopping environment. The top surface of small hall is set with circular LED light boxes serving as guidance. Diamond-shaped LED lamp beads are built in the light box to make the box look like starry sky at night through mirror refraction.

10

11

椭圆中庭，围绕观光电梯周围的侧裙板采用方格纹理的发光玻璃，透露出晶莹剔透，美轮美奂的艺术效果；大面积侧裙板通过体块的穿插，以及漂浮设计手法，营造出轻盈灵动的空间特色；地面时尚动感的拼花图案打破了地面呆板沉闷，图案犹如交错的光晕，尽显华贵典雅的气质。

Side skirt boards in the oval atrium and around panoramic lift are made of luminescent glass in square pattern, crystal clear and amazing. Interspersed large side skirt boards and floating design method create a light and smart space. Fashionable and dynamic floor patterns break the sense of stiffness and oppression, and such pattern is of elegant quality just like interlaced halos.

12

13

圆中庭采用了体块穿插的手法，两种颜色、质地强烈对比；直线体块在空间中穿插；以及灯光营造的视觉错位效果等，使圆中庭的视觉进深感加强，从而产生出较大的视觉空间感。

The circular atrium adopts the method of block intersection. Strong contrast between two colors and textures, intersection of straight line blocks in the space and a visual dislocation effect realized through light design jointly enhance visual depth of the circular atrium, thus creating a good sense of visual space.

10 椭圆中庭剖面图
11 椭圆中庭
12 圆中庭
13 圆中庭俯视

15

16

在直街空间中，运用了平铺直叙的表现手法，强调侧裙板的肌理对比和色彩的搭配，侧裙铝板材质细微的凹凸感和精致的工艺收口处理，令空间具有很高的细节品质；同时，在室内空间中首次使用"模块化"设计，将外露的设备末端与顶面造型，侧裙板的工艺分缝及地面拼花在一定的比例模数中进行设计与排列，使长街看起来井然有序，产生了强烈的视觉效果。

The straight gallery design is plain and un-garnished, stressing side plate texture comparison and color matching. Subtle concave-convex feeling of aluminum plates on side skirt boards and good binding process enhance detail quality of the space. Meanwhile, "modular" design is adopted in interior space design for the first time. Exposed equipment terminal, top shape, process parting of lateral wall and floor patterns are designed and arranged in moduli as per a certain proportion, showing people an orderly street of strong visual effect.

14 室内步行街连桥
15 自动扶梯
16 室内步行街侧裙板

17

LANDSCAPE OF PLAZA
广场景观

景观的主题是“灯影秦淮”，用现代手法演绎秦淮河两岸景观，使人仿佛重游秦淮河，在画舫凌波中，来一场时空之旅，阅览秦淮河的历史文化。

The landscape design of the Plaza is themed by "Qinhuai in light shadow", depicting landscapes on both banks of Qinhuai River by modern means and making one feel as if revisiting the River, reading history and culture of the River in gaily-painted pleasure-boat and glittering water.

18a

17 广场主雕塑
18 景观灯笼
19 广场水景

18b

18c

19

万达广场
WANDA PLAZA
万达广场
永辉超市

NIGHTSCAPE OF PLAZA
广场夜景

在夜景照明设计中，结合其建筑形式特征——扭曲、折叠、错位的视觉效果——呈现灯光的曲线美、律动美、和谐美。在照明设备的选用上，尽量采用技术成熟、稳定可靠的通用标准型产品，尽量避免非标制品的大量使用。这样既便于有效地控制工程造价，又便于施工、调试和后期的长效运用。

Based on characteristics of architectural form - twisty, folding and malposed visual effect, the nightscape design of the Plaza showcases beauty of curve, rhythm and harmony of lights. Commonly used standard lighting facilities which are mature in technology and stable & reliable in performance are selected as much as possible. Meanwhile, understandard products are refrained from being utilized in large quantity. These operations not only effectively controls engineering cost but also facilitate construction, debugging and later-stage utilization.

21

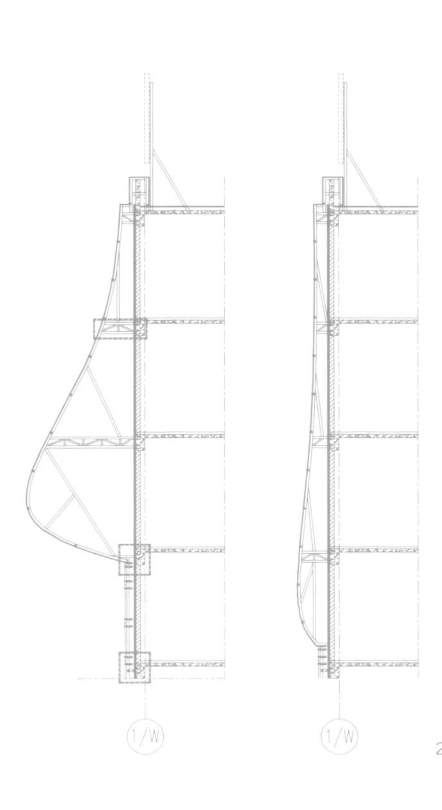

22

20 广场全景
21 门头夜景
22 入口灯具安装示意图

IMAX

23 二号入口夜景

OUTDOOR PEDESTRIAN STREET
室外步行街

室外步行街内采用当地的文化，提取最具特色的图案及单体进行艺术处理，增添文化气氛。挑选南京古都最具代表性的唐、明、清及民国时期经典文化符号进行艺术处理，在室外步行街内形成一条时间轴，给商业氛围增添文化内涵。

The exterior pedestrian street design extracts the most distinctive patterns and individuality in local culture for artistic treatment so as to enhance cultural atmosphere. Through artistic treatment of the most representative classic cultural symbols of Nanjing during the Tang, Ming and Qing Dynasties and the period of the Republic of China, a time line is formed in the exterior pedestrian street to enhance cultural connotation to the commercial environment.

25

26

24 室外步行街外立面
25 室外步行街转角
26 室外步行街店铺

27a

27b

28a

28b

28c

28d

27 室外步行街外立面夜景
28 室外步行街景观小品

29

30

FACADE OF HOTEL
酒店外装

酒店是"L"形平面酒店，石材造型以精细的竖纹体现奢华与厚重，衬托出江宁的历史文化底蕴。

The hotel is in an "L" shape in layout, adopting stones with fine vertical stripes to reflect luxury and massiveness as well as historical and cultural deposits of Jiangning.

31

29 酒店外立面
30 酒店入口
31 酒店夜景

 酒店入口雨棚

NIGHTSCAPE OF HOTEL
酒店夜景

嘉华酒店夜景设计旨在营造出一种低调内敛、温馨典雅的夜间形象；用间接光照明方式表现出建筑的质感和肌理。在光色设计上运用了整体暖黄色调，集功能性照明与装饰性照明于一体。灯光设计旨在突出入口，体现出近人尺度建筑自身的细节，高端、精致、典雅。

The nightscape of Wanda Realm Hotel Nanchang is designed to create a humble, warm and elegant image at night by using indirect illumination to reflect the fabric and texture of the building. The color of warm yellow, which combines functional lighting and decorative lighting, has been applied in the aspect of light & color design. The lighting design is to emphasize the entrance to reflect the details of the building with the sense of human scale, presenting its exquisite, delicate and elegant image.

NIGHTSCAPE OF HOTEL
酒店景观

酒店后庭院的水中漂浮的室外木平台及一丛丛修竹，营造静谧禅意的空间。

The hotel back courtyard is set with an outdoor platform floating on the water surface, combined with tall bamboos, creating a tranquil space full of Buddhist mood.

33

34

35

36

33 酒店后庭院夜景
34 后庭院水景
35 喷泉
36 后庭院小径

CHONGQING WANZHOU WANDA PLAZA
重庆万州万达广场

开业时间 2013 / 07 / 05
建设地点 重庆
占地面积 6.25 公顷
建筑面积 21.38 万平方米

OPENED ON JULY 5 / 2013
LOCATION CHONGQING
LAND AREA 6.25 HECTARES
FLOOR AREA 213,800 m²

2

OVERVIEW OF PLAZA
广场概述

重庆万州万达广场总建筑面积21.38万平方米，总投资逾20亿元，由大型购物中心、商业步行街、高级影院、希尔顿逸林酒店组成，其中购物中心建筑面积达14万平方米；引入万达百货、万达影城、大歌星KTV、大玩家、苏宁电器、永辉超市六大主力店，H&M快时尚品牌、NOVO潮流集合店。室内精品街引进CK JEANS、TOMMY HILFIGER、GANT、FILA、HAZZYS、JORYA、APPLE等国际知名品牌，有66个品牌为首次进驻万州市场。

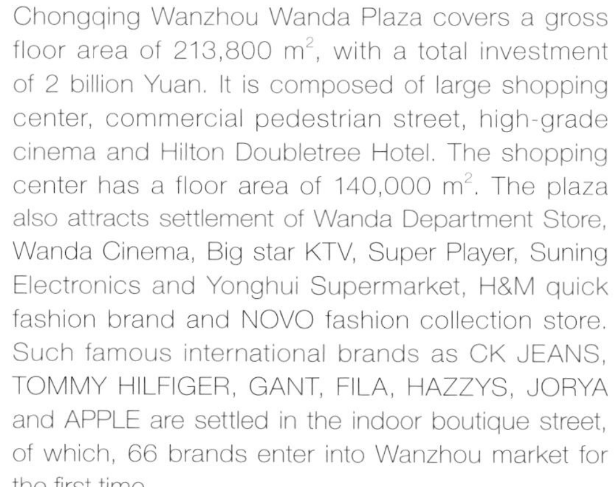

Chongqing Wanzhou Wanda Plaza covers a gross floor area of 213,800 m², with a total investment of 2 billion Yuan. It is composed of large shopping center, commercial pedestrian street, high-grade cinema and Hilton Doubletree Hotel. The shopping center has a floor area of 140,000 m². The plaza also attracts settlement of Wanda Department Store, Wanda Cinema, Big star KTV, Super Player, Suning Electronics and Yonghui Supermarket, H&M quick fashion brand and NOVO fashion collection store. Such famous international brands as CK JEANS, TOMMY HILFIGER, GANT, FILA, HAZZYS, JORYA and APPLE are settled in the indoor boutique street, of which, 66 brands enter into Wanzhou market for the first time.

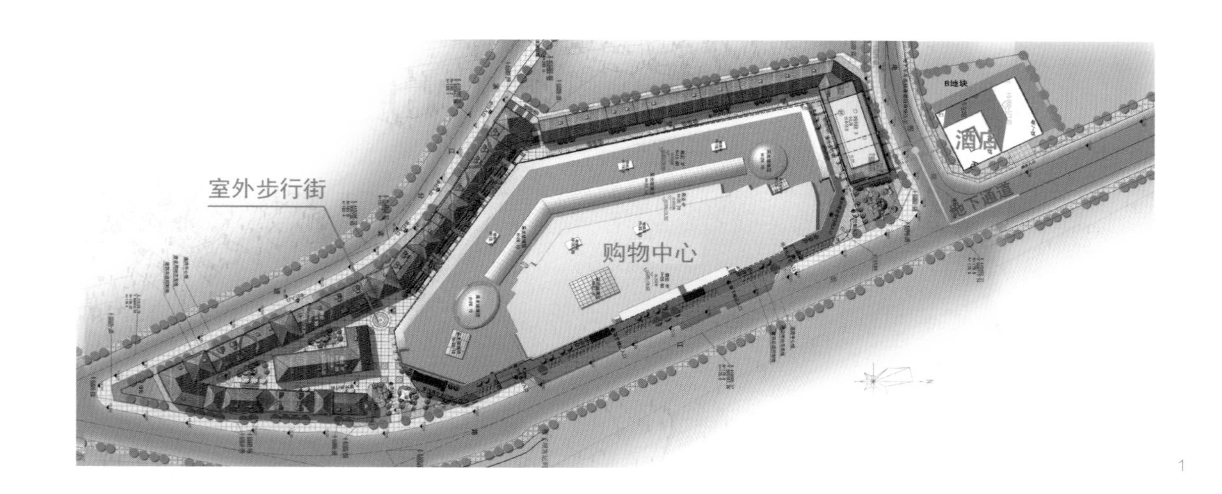

1

1 广场总平面图
2 广场外立面
3 广场外立面效果图

3

FACADE OF PLAZA
广场外装

重庆万州万达广场外立面装饰风格折射了近代Art-Deco装饰艺术派的简化欧式风格。作为整体商业风格中西合璧主题的一部分，以石材、幕墙、黑镜钢、LED大屏等现代建筑元素，欧式立面构成手法组合，诠释了充满现代感的商业气质。

The façade decoration reflects simplified European style of the Art-Deco decorative art school in modern times. The overall commercial style is themed by a combination of Chinese and western elements. As part of this, modern architectural elements including stone, curtain wall, black mirror steel, LED large screen are adopted. Meanwhile, European elevation composition method are utilized to interpret modern commercial feature of the Plaza.

4 广场一号入口
5 广场立面图
6 广场二号入口

4

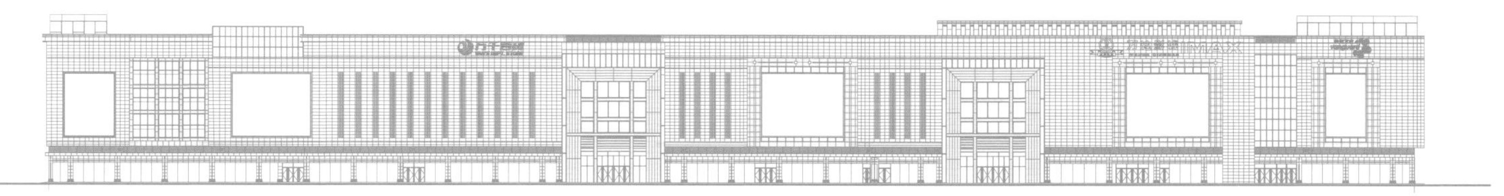

5

6

购物中心临街面通过立面构图单元富有韵律感的排列，通过横向与竖向的段落划分，演绎着欧式风格的严谨与丰富。

In designing frontage of the shopping center, elevation composition units are arranged rhythmically, and horizontal & vertical section division are utilized, thus showcasing preciseness and abundance of the European style.

7

8

9

7 百货外立面
8 广场一号入口
9 广场二号入口

作为万州港门前的万州城市名片，南侧门头突出的建筑形象与民国风情步行街入口的标志塔相映成趣，中西合璧的商业风格创新演绎得淋漓尽致。

As the name card of Wanzhou City, the southern gate in front of Wanzhou port highlights an architectural image which corresponds to landmark tower at the entrance of pedestrian street with features of the Republic of China. It incisively and vividly interprets commercial style innovation featured by a combination of Chinese and western elements.

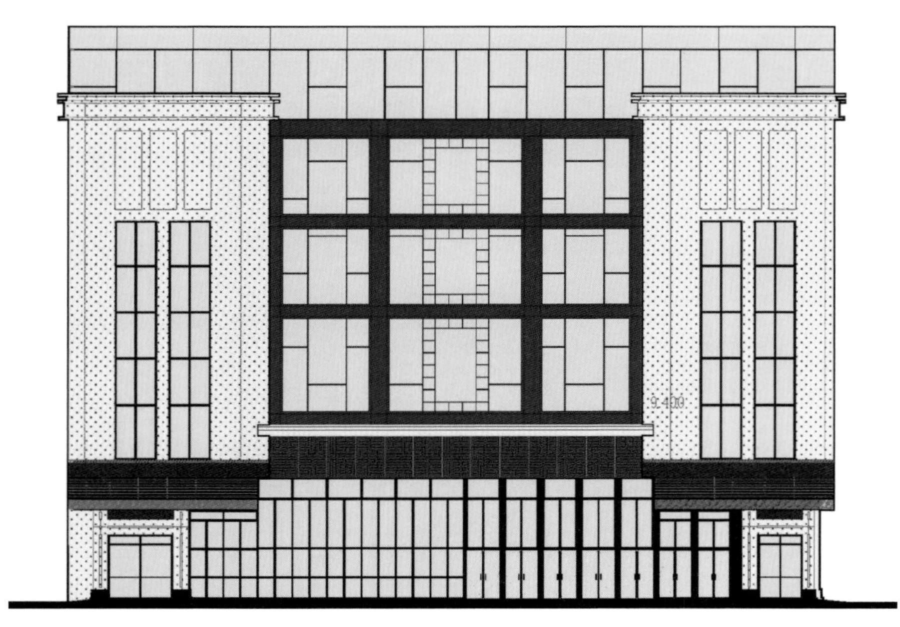

10

11

12

13

门头设计将传统的幕墙单元与LED发光模组相结合，巧妙地兼顾了浑然一体的造型设计和异形屏幕的媒体展示功能，更成为三峡城市中一颗璀璨的夜明珠。

In the gate design, traditional curtain wall unit and LED luminescent module are combined to skillfully balance integrated shape design and media display function of irregular screen, making the project look like a dazzling pearl in cities in the Three Gorges area.

10 大商业立面图
11 广场四号入口
12 广场三号入口
13 广场二号入口

14

15

INTERIOR OF PLAZA
广场内装

内装的设计自然而然的也以“江城”这一概念作为设计的创意原点。自然和生命是核心的设计理念，也只有这样才能体现出这里的特色。在设计手法上通过对空间形体的再塑造，运用体块的穿插、材质的变换体现出江与岸、水与城的相辅相成，紧密融合的关系。流线的具有宽窄变化的线形元素如滚滚长江，成为贯穿整个购物空间的典型符号。侧裙造型，观光梯的图案，门厅的吊顶均由此生发而来。

14 圆中庭
15 圆中庭侧裙板

The interior design is based on the "River City" concept. Nature and life are core design concepts, and only these concepts can truly reflect characteristic of the river city. Re-shaping of space form, blocks interspersion and material changes jointly create a complementary and closely integrated relationship between river and bank and between water and city. Linear elements with variations in width look like surging Yangtze River, becoming a typical symbol of the whole shopping space. It is also a source of inspiration in designing lateral wall shape, panoramic lift patterns and suspended ceiling in lobby.

16

椭圆中庭的侧裙，通过GRG材料的运用，产生凹凸起伏的体块变化，再通过不同色彩的运用，更加强调了它们的块面结构。宽窄不一的线形，每一层不同的穿插关系让空间整体而又富有变化，使江与岸的关系更加贴切而生动。

Lateral wall in the oval atrium is made of GRG material, with concave-convex block variations, added with utilization of different colors, further stressing their block structure. Lines of different widths and different intersperse relationships on different layers create an integrated but variable space, and form a close and vivid relationship between river and bank.

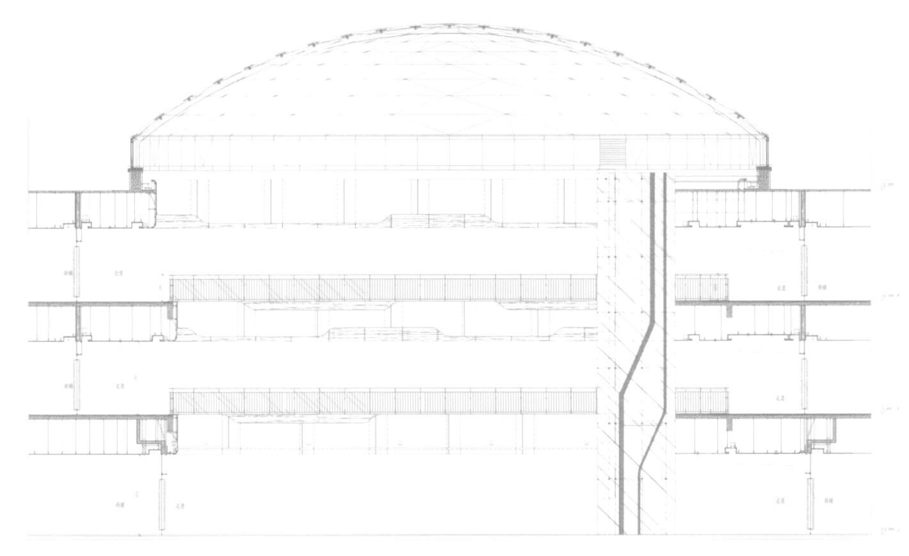

17

16 椭圆中庭
17 椭圆中庭剖面图
18 采光顶
19 圆中庭

18

圆中庭观光电梯上的线形条纹图案的运用，让观光电梯犹如被线条分割的不同航道，又像是气势雄浑的瀑布飞流而下。

Panoramic lifts in the circular atrium are designed with linear strip patterns, making the panoramic lift appear like different channels divided by lines and vigorous waterfall plunging down.

19

20 椭圆

WIGU
Calvin K
ONLY

22

直街的设计，将二、三层的天桥设计成交错叠加的关系，产生更加丰富而有层次的共享空间，改变了原建筑天桥相对呆板的形象，空间的品质也因此得以提升。侧裙和天桥的造型采用统一的处理手法，运用条形的木纹铝板和白色GRG在颜色和质感上形成对比，让天桥成为两种材质天然的交汇节点，将人的视线自然引向了天桥这个长街空间中的视觉中心。这种处理巧妙而含蓄，也让长街空间更为和谐统一。

In the straight gallery design, overpasses on the second and third floors are staggered and overlaid to create a rich and hierarchical shared space, thus changing stiff image of the original building overpass and enhancing the space quality. Lateral wall and overpass shape are uniformly designed, utilizing strip-type wood grain aluminum plates and white GRG to form a contrast in color and texture, making overpass a natural node where two materials meet, and naturally guiding pedestrian's sight line to the overpass, the visual center of the long gallery space. Such skillful and implicit treatment method makes the long gallery space more harmonious.

23

21 室内步行街
22 连桥
23 连桥设计手稿

24

LANDSCAPE OF PLAZA
广场景观

重庆万州万达广场景观设计通过基本单元的组合连接，营造出既有生活气息又符合经营要求的广场与街巷空间。

In landscape design of Chongqing Wanzhou Wanda Plaza, the combination and connection of basic units create a plaza and street space which is not only alive but also satisfies operation requirements.

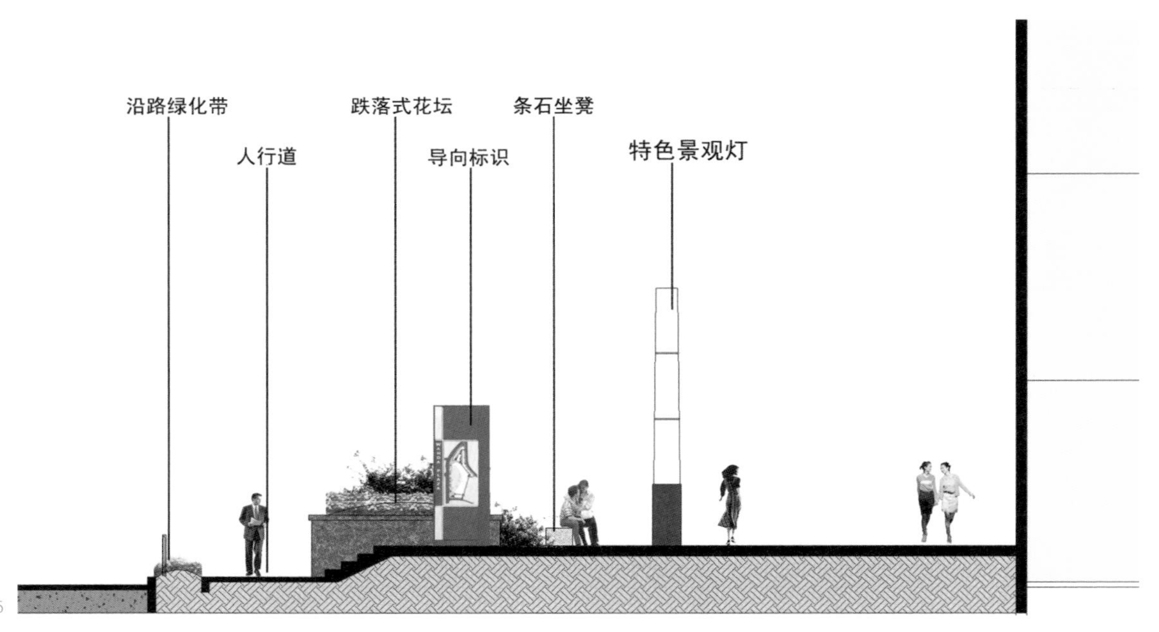

25

24 广场座椅
25 广场景观剖面图
26 广场喷泉
27 广场花坛

①
街
26

NIGHTSCAPE OF PLAZA
广场夜景

万州万达广场，犹如一条城市之舟、商业之舟，坐落在长江之畔。建筑不仅仅是把各种美丽的花岗石或大理石堆叠起来，更是把最具现代化的购物、休闲、娱乐、生活氛围带到了重庆人民的生活中，形成了一个真正意义上的城市中心。在灯光装扮下，这艘“巨轮”仿佛航行在夜空中的银河里，让人们产生无限的遐想。整体照明以暖色调为主，重点入口部分增加了动态光，达到吸引人流，提高关注度的效果，烘托商业气氛，成为夜晚的焦点场所。

The Plaza appears like a boat of city and a boat of commerce situated at the bank of Yangtze River. Architectural design is not merely a stacking of various beautiful granites or marbles; rather, it is a process of integrating modern shopping, leisure, entertainment and living functions into daily life of people in Chongqing, with the aim to form an urban center in the real sense. The lighting design creates an image that the "giant ship" is navigating in the Milky Way at night, making people be lost in wild and fanciful thoughts. The overall lighting adopts warm colors. The dynamic lights are added at key entrance to attract pedestrian flow, enhance attention, reflect commercial atmosphere and serve as a focal place at night.

28 二号入口夜景
29 一号入口夜景

28

万达广场®
WANDA PLAZA
永辉超市
H&M
万州金街
50
①

30

31

30 五号入口夜景
31 四号入口夜景
32 外立面夜景
33 二号入口夜景

32

33

万州万达广场中西合璧的风格构成中，在建筑入口处，整体形成中西合璧的建筑景观。夜景照明设计充分考虑到这些要素，通过光影、光色把建筑立面丰富的材质表现得富有韵律。

The Plaza style is featured by a blending of Chinese and western elements. The building entrance landscape design reflects this style. Nightscape design, by considering this element, attempts to express a rhythmic building elevation with virtue of light & shadow and light color.

OUTDOOR PEDESTRIAN STREET
室外步行街

在建筑方案中虽然采用了传统的民国建筑形式及简欧建筑形式，景观设计却没有一味附和这些形式，而是采用多视角的原则，具备创新和时代感。同时符合了万达广场的商业氛围和灵气所在，没有拘泥于某一时期的某一特定风格，而是建成风格独特的、在重庆代表万州风情的、在中国体现重庆特色的、在中国体现时尚的商业街，这才是本方案广场景观设计宗旨。

The architectural scheme adopts traditional form of buildings constructed during the Republic of China and simplified European architectural style, but landscape design is not a simple repetition of these styles; rather, it adopts multi-perspective principle, being innovative and modern, in line with commercial atmosphere and anima of Wanda Plaza. Instead of adopting a certain style during a certain period of time, it attempts to construct a commercid street which reflects characteristics of Wanzhou in Chongqing, reflects characteristic of Chongqing in China and reflects fashion in China. This is the design purpose of the landscape design.

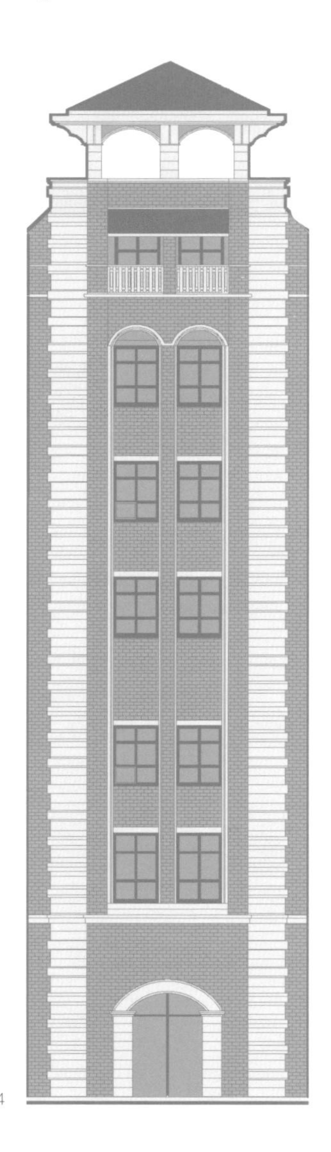

34

35

34 室外商业街钟楼
35 室外商业街外立面

SUNSHINE
倾城绽放
2013.7.5

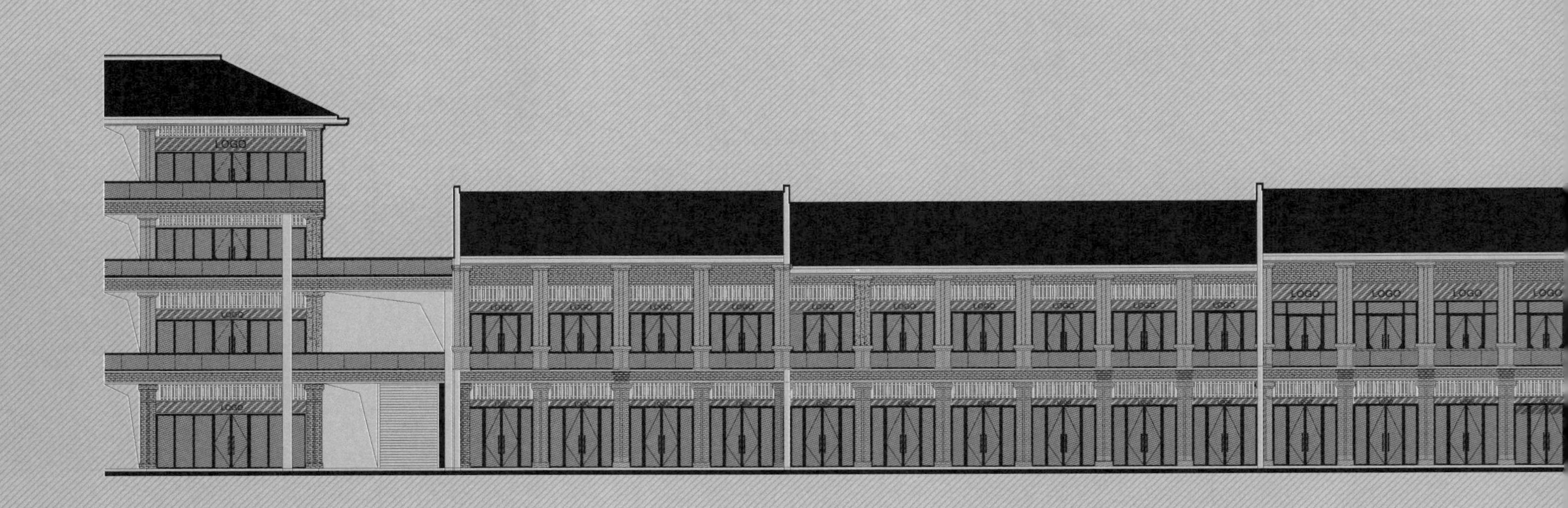

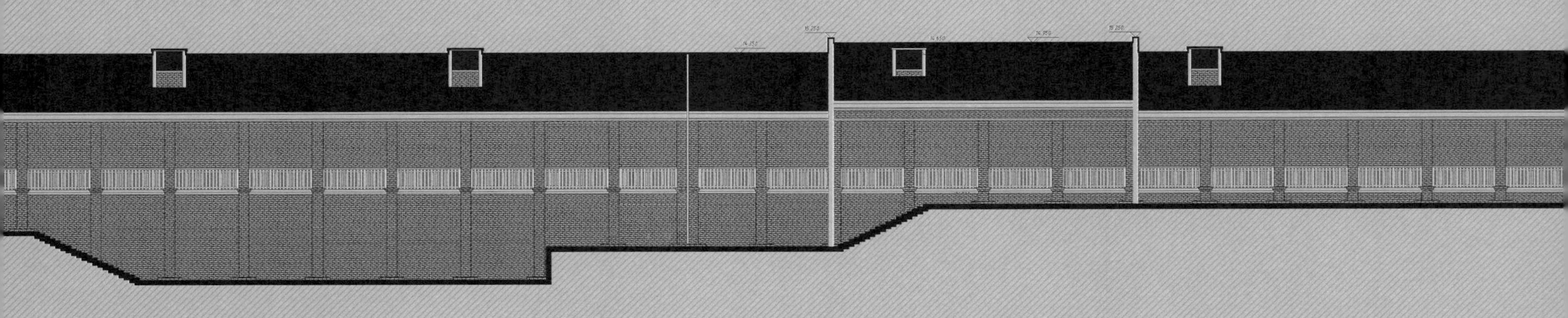

36 室外步行街立面图

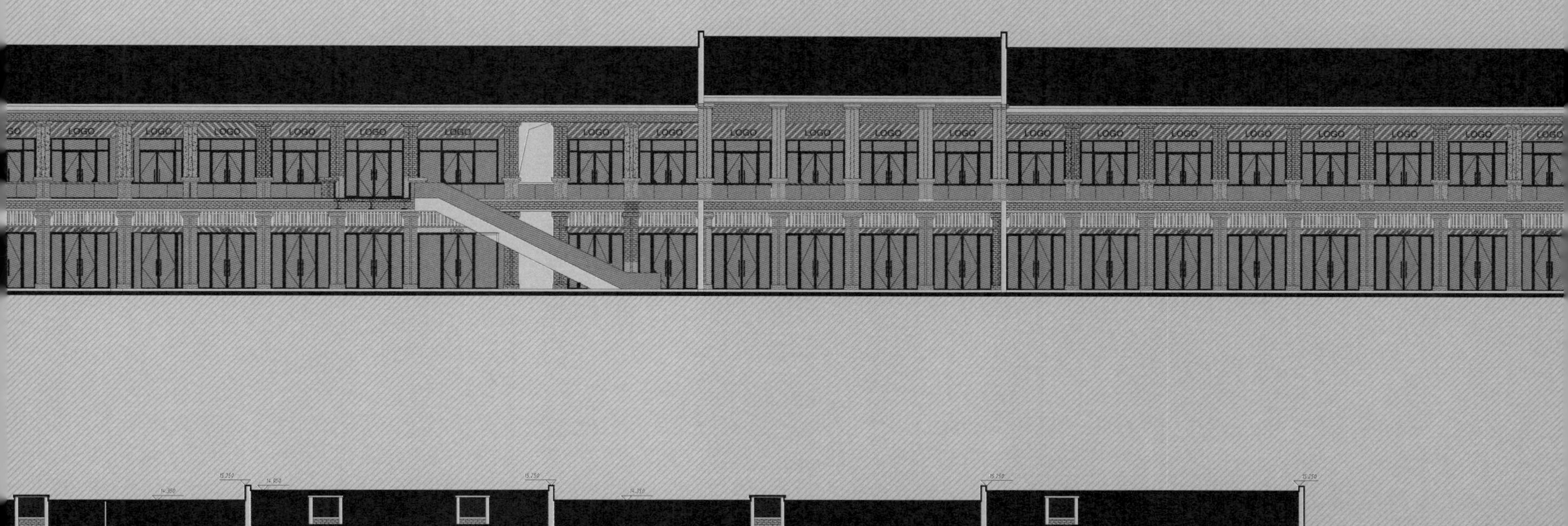

Shakespeare
倾城绽放
2013.7.5
ONBKO
GOLNUBIAS
Wonderful

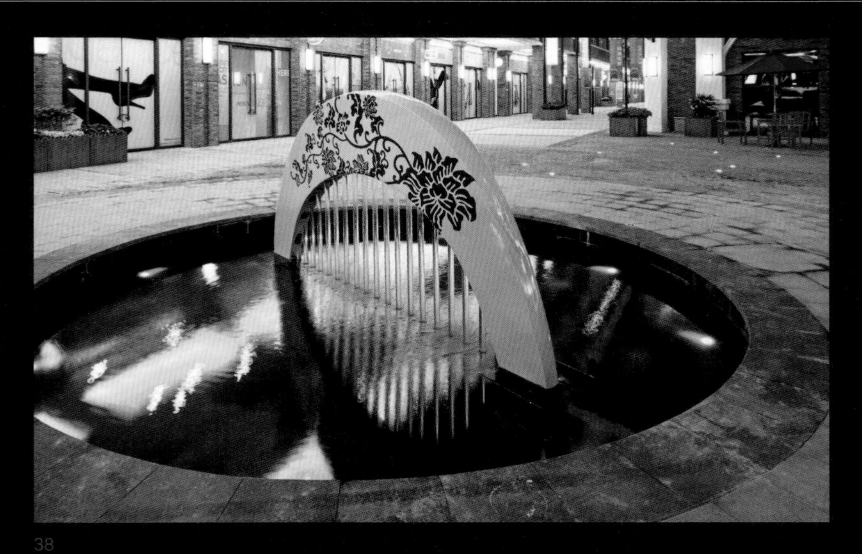

38

39

40

37 室外步行街夜景
38 水景
39 喷泉
40 景观小品

FACADE OF HOTEL
酒店外装

重庆万州希尔顿逸林酒店从贯穿万州的生命之流长江中吸取灵感，选择变化万千的水纹和水岸形态作为酒店的设计元素。通过对传统文化的提炼和升华来展现五星级酒店的奢华与文化内涵。酒店总建筑面积3.64万平方米，公共区总面积1.24万平方米，客房总面积1.73万平方米，后勤总面积0.67万平方米。建筑层数21层，地上19层，地下2层，建筑总高度86.30米，是渝东地区首家国际五星级酒店。

Chongqing Wanzhou Hilton Doubletree Hotel design is inspired from the Yangtze River, a river of life running through Wanzhou. Diverse water waves and water edge forms are selected as design elements. Traditional culture is extracted and refined to reflect luxury of the five-star hotel and its cultural connotation. The hotel has a gross floor area of 36,400 m^2, including 12,400 m^2 public area, 17,300 m^2 guest room area, and 6,700 m^2 logistics area. The hotel has 21 stories, including 19 stories above ground and 2 stories underground. The 86.30 m high building is the first international five-star hotel in east Chongqing district.

41

42

41 酒店外立面
42 酒店总平面图
43 酒店夜景

DOUBLETREE
万达希尔顿逸林酒店
万达广场®
WANDA PLAZA

LANDSCAPE OF HOTEL
酒店景观

在酒店的景观设计中，充分尊重现状地块，并与相邻的大商业广场形成呼应，在大环境的营造中既保持酒店自己独特的景观定位，又继承了大商业的诸多设计亮点。景观设计采用多视角的原则加以创新并表现时代感，打造代表万州风情的、体现重庆特色的、具有时尚感的酒店景观。

The hotel landscape design, by considering the existing land lot, corresponds to its adjacent large commercial plaza, maintaining its distinctive landscape orientation in large environment creation while demonstrating many highlights in large commercial center design. The landscape design is a kind of innovation based on multiple-perspective principle with the aim to reflect a sense of the times and to create a modern Landscape of hotel representing characteristics of Wanzhou and Chongqing.

44

45

46

44 酒店绿化
45 酒店水景
46 酒店入口景观

万达希尔顿逸林
DOUBLETREE
BY HILTON

47

47 酒店入口夜景
48 酒店外立面夜景

NIGHTSCAPE OF HOTEL
酒店夜景

酒店的灯光设计融合了现代建筑灯光大气和传统商业灯光的热烈。整体照明方式以体现建筑的线条感为主，在统一了照明手法的基础上又有一些动态光的变化，使整个建筑既稳重又不失韵律感。通过智能灯光控制系统可以在运营过程中根据需要选择合适的照明效果。

The hotel light design is a combination of grand modern architectural light and traditional commercial light elements. The overall lighting reflects a sense of line of the building, with variations of dynamic lights based on unified lighting, thus making the whole building modest and rhythmical. Besides, intelligent light control system enables appropriate lighting effect during operation.

审图号：GS（2014）1915号

PART

C WANDA COMMERCIAL PLANNING RESORT HOTELS 度假酒店

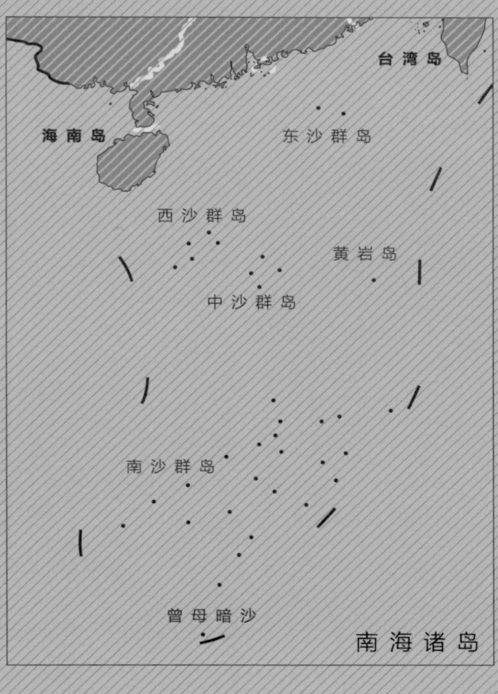

PARK HYATT CHANGBAISHAN
长白山柏悦酒店

开业时间 2013 / 09 / 15
建设地点 吉林 / 长白山
客房数量 163 间
建筑面积 3.49 万平方米

OPENED ON SEPTEMBER 15 / 2013
LOCATION CHANGBAISHAN / JILIN PROVINCE
GUEST ROOMS 163
FLOOR AREA 34,900 m^2

OVERVIEW OF HOTEL
酒店概述

长白山柏悦酒店位于中国吉林省长白山市抚松县万达长白山国际度假区南区，毗邻长白山凯悦酒店，总规划用地5.55公顷，酒店建筑面积3.49万平方米，163把钥匙，其中包括36间套房及9栋别墅。大多数客房均拥有私人阳台，夏季一览长白山秀丽景色，冬季可尽享滑雪场大美山景。酒店内豪华水疗、健身设施及室内泳池一应俱全，为宾客提供一系列身心放松的疗程和健身空间。宴会厅拥有260平方米的会议空间及3个小会议室。

酒店定位为滑雪度假酒店，直面长白山滑雪场的主雪道，滑雪住客可以直接滑进滑出。大堂及客房均可以欣赏到壮丽的雪道景色和周围山景。整体建筑风格源于北美正统的庄园，自然的坡顶、舒展的平面，完美贴合大自然。采用经典建筑元素符号，如气派的门廊、千姿百态的屋顶、画龙点睛的烟囱，创新地诠释了“世外桃源”体验式的奢华酒店。

1

1 酒店总平面图
2 酒店鸟瞰图

2

Adjacent to Hyatt Regency Changbaishan, Park Hyatt Changbaishan is located in the southern district of Wanda Changbaishan International Resort, Fusong County in Changbaishan, Jilin Province, occupying total planned land of 5.55 hectares and floor area of 34,900 m^2. Among the 163 guestrooms, 36 suites and 9 villas are included. A majority of guestrooms feature private verandas, enabling an overview of Changbaishan beauty in summer and mountain beauty of ski in winter. The luxury spas, fitness facilities and indoor swimming pools available in the hotel provide guests with a series of relaxation medications and fitness space. The banquet hall houses 260 m^2 meeting space and three small meeting rooms.

Positioned as a ski resort, the hotel faces the main ski track of Changbaishan Ski Resort, enabling the direct access of ski guests and appreciation of magnificent ski track scenery and surrounding mountains from lobby and guestrooms. The overall architectural style-natural slope and stretching plane perfectly fitting the nature, is originated from the orthodox manor in North America. Meanwhile, the classic architectural element symbols, such as the gorgeous vestibule, oddly-shaped roof and finishing touch-chimney, interpret a "Land of Idyllic Beauty" experiential luxury hotel in an innovative manner.

FACADE OF HOTEL
酒店外装

外立面整体设计理念：建筑因山就势，外立面下部采用了大量石材，与周边地貌高度契合，展现了独特的山地地域特色。主体建筑控制在3到4层，而别墅类附属建筑的层数则控制在2层，建筑轮廓与起伏的地形、林际线相呼应，形成丰富的天际线。立面石材采用天然碎拼手法，结合多变的间色及凹凸的质感，形成自然的生长气氛和个性。

Overall design concept of façade: nestled among mountains, the building's façade bottom uses a large quantity of stones to highly match the surrounding landscape, presenting unique mountain regional features. The main building is confined to three to four-storey while auxiliary building like villa to two. The building outline, echoing with the undulating terrain and forest boundary line, forms a rich skyline. Via applying naturally fragmented piecing method and integrating variable secondary colors and bumpy texture, the façade stones present natural growth atmosphere and personality.

3

4

5

3 酒店外立面
4 酒店别墅区效果图
5 酒店立面图

6 酒店入口

7

7 酒店屋顶
8 酒店外立面
9 屋顶烟囱

8

9

11a

11b

11c

10 酒店入口雨棚
11 入口大门及门套设计优化前后对比

12

12 酒店景观
13 户外餐饮平台

LANDSCAPE OF HOTEL
酒店景观

景观设计提供更多宜人的私密小品空间，营造极致的浪漫氛围和富有情趣的六星级酒店享受：如在户外餐饮平台部分设置了一个独一无二的花园式私人用餐空间。

柳暗花明又一村——独特的到达体验。客户会首先由主路右转穿过一个密林景观带，然后穿过一条古朴石桥，再迂回到酒店的大堂。空间层层叠叠，又豁然开朗。

The landscape design intends to render more pleasant private accessories space, and create an extremely romantic atmosphere and a feast of six-star hotel full of fun, e. g. A unique garden-style private dinning space is available at outdoor F&B platform.

Every cloud has a silver lining-unique access experience. Guests will firstly turn right from the main road to pass through a jungle landscape and an ancient stone bridge, and then detour to the hotel lobby with a sense of intercrossed while suddenly extensive space.

14 酒店景观雪景

NIGHTSCAPE OF HOTEL
酒店夜景

长白山柏悦酒店的照明设计，旨在提供一个非传统的、倾向于田园风光的设计概念。烟囱的照明与长白山凯悦酒店相类似，意在展现屋顶边缘的照明，并且在两个酒店间提供统一的视觉效果。檐下淡淡地勾勒出暖色的光，创造统一的视觉感受，体现建筑顶部结构之美。同时对石“柱”支撑进行洗墙照明，渲染纯欧式滑雪小镇。建筑以木、石材料为主的原生态、回归天然的审美意境。外立面木柱柱身的凹槽造型，设置向上的小投光灯，强调建筑的结构的纤巧之美，营造灵动活跃的气氛。

The lighting design of Park Hyatt Changbaishan intends to interpret a non-traditional and idyllic scenery-oriented design concept: similar to Hyatt Regency Changbaishan, the lighting of chimney aims to highlight the roof edge and build a unified visual effect between two hotels; warm color light is transmitted under eaves to also present a unified visual effect and showcase the beauty of building's top structure. Flood light is applied to stone "Column" support, hoping to embellish original ecology centering on wood and stone, and aesthetic conception of returning to the nature characterizing pure European ski town buildings; the groove shape of façade wood column body is furnished with small spotlights upward to emphasize the delicacy of architectural structure and build a free and lively atmosphere.

15 酒店景观夜景

16

16 酒店屋檐
17 酒店外立面
18 烟囱特写
19 酒店导向标识
20 景观灯

17

18

19a

19b

20

HYATT REGENCY CHANGBAISHAN
长白山凯悦酒店

开业时间 2013 / 09 / 15
建设地点 吉林 / 长白山
客房数量 278 间
建筑面积 5.51 万平方米

OPENED ON SEPTEMBER 15 / 2013
LOCATION CHANGBAISHAN / JILIN PROVINCE
GUEST ROOMS 278
FLOOR AREA 55,100 m²

OVERVIEW OF HOTEL
酒店概述

长白山凯悦酒店位于中国吉林省长白山市抚松县万达长白山国际度假区南区，毗邻度假区商业街，总规划用地5.74公顷，酒店建筑面积5.51万平方米，278把钥匙，其中包括33间豪华套房。大多数客房均拥有私人阳台，夏季长白山秀丽景色尽收眼底，冬季滑雪场山景一览无余。酒店里餐厅、酒吧、茶肆、咖啡等餐饮空间就有8处之多，同时泳池、健身、SPA、户外温泉、滑雪服务等康体配置一应俱全，还特别为小朋友设置了儿童活动中心，提供丰富的活动体验。作为度假区的会议酒店，拥有1000平方米的大型宴会厅及多达9个中小会议室，会议设施达到国际一流。

作为滑雪度假酒店，其对面即为长白山滑雪场的主雪道，滑雪住客可以方便地直接滑进滑出。大堂及客房均可以欣赏到壮丽的雪道景色和周围山景。整体建筑风格以充满北美风情的山野村居为灵感，将典型美式乡村风格同长白山独特的地域文化特征巧妙结合，同时融入大量的文化元素。

Adjacent to Commercial Street of resort, Hyatt Regency Changbaishan is located in the southern district of Wanda Changbaishan International Resort, Fusong County in Changbaishan, Jilin Province, occupying total planned land use of 5.74 hectares and floor area of 55,100 m². Among the 278 guestrooms, 33 luxury suites are included, and a majority of guestrooms feature private verandas, enabling an overview of Changbaishan beauty in summer and mountain beauty of ski in winter. The hotel boasts of up to eight dinning space covering restaurants, bars, teahouses, cafeterias, available sports configuration such as swimming pool, fitness facilities, spa, outdoor hot spring and ski service and Children's activity center specially made for kids to render rich activity experience. Besides, as a meeting hotel of resort, Wanda Hyatt Regency Changbaishan, equipped with world-leading meeting facilities, consists of a grand banquet hall of 1,000 m² and up to nine middle & small meeting rooms.

As a ski resort, the hotel faces the main ski track of Changbaishan Ski Resort, enabling the direct access of ski guests and appreciation of magnificent ski track scenery and surrounding mountains from lobby and guestrooms. Inspired by the village scene with North America flair, the overall architectural style skillfully integrates the typical American country style with unique Changbaishan cultural traits, and meanwhile, infuses plentiful cultural elements.

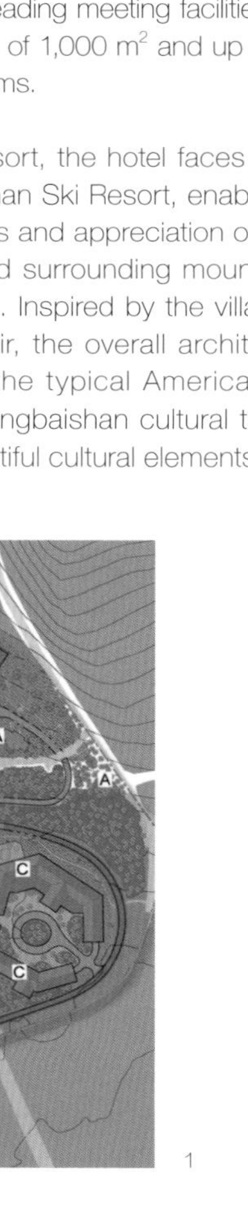

1

1 酒店总平面图
2 酒店鸟瞰
3 酒店立面图

2

3

4

5

4 酒店全景图
5 酒店主入口雪景

FACADE OF HOTEL
酒店外装

外立面整体设计理念：建筑低缓的人字形屋面，使檐口的流线更加流畅；深檐下暴露的椽头和装饰木托架，力求贴近自然的风格，使建筑与周边森林环境有机地融为一体。建筑材料以木、石材料为主，回归原生态、天然的审美意境。外立面采用大量石材，与周边自然地貌高度结合，将北美建筑特色与传统中国山地建筑风格完美结合。

Overall design concept of façade: lower gable roof makes a smoother streamline of cornice; the nature-like pursuit of exposed sally and decorative wooden tray under deep eave make the building and forests surrounded organically integrated. The use of wood and stone as the main building materials gives birth to original ecology and natural aesthetic conception, and the use of a large quantity of stones in façade highly matches the surrounding landscape and perfectly integrates architectural features of North America with mountain building style of China as well.

7

8a

8b

6 酒店入口雨棚
7 酒店鸟瞰
8 屋顶特写

酒店主入口门头上增放石材线角，使建筑成为一个有机的整体。设计管控中，通过对建筑细部的把握和刻画，提升了整体品质。

The addition of stone line-angle at main entrance gate of hotel makes the building an organic whole. By grasping and depicting the architectural details in the process of design control, the overall quality is enhanced accordingly.

9

9 酒店入口设计优化前
10 酒店入口门头设计优化后

10

11

LANDSCAPE OF HOTEL
酒店景观

景观设计理念是打造出一个配合得天独厚的自然环境、四季皆宜的滑雪度假区酒店氛围。景观设计把酒店建筑的内外空间完全无缝地融入森林和山体大自然的环境中，酒店外围景观是欣赏自然，内部景观则是用精细的人工雕琢来诠释自然。

In order to interpret the concept of creating an ideal ski resort hotel atmosphere for four seasons and in coordination with unique natural environment, the landscape design seamlessly incorporates the inner and outer space of hotel into the nature hugging by forests and mountains-appreciating nature through external landscape and interpreting nature through fine artificially carved internal landscape.

12

11 酒店鸟瞰
12 骑楼的“月亮门”造型

13a

13b

14

13 景观小品
14 景观石板路
15 酒店雪景

15

NIGHTSCAPE OF HOTEL
酒店夜景

长白山凯悦酒店的照明设计旨在加强建筑设计感并提供一个强有力的视觉效果，以突显其建筑形象。房间的内透光是有故事的光，是夜景照明的重要组成语言，传达温暖小镇的理念。照明重点是传统的“阿尔卑斯”桁架屋顶结构，同时突出阳台及其结构支撑，将整个建筑楼宇融入环境中。通过阳台的投影，对建筑周边石材进行洗墙照明，进一步加强与整个园林的连接。

The lighting design of Hyatt Regency Changbaishan intends to reinforce the sense of architectural design and a powerful visual effect, highlighting its architectural image. Through the story-telling interior lights, also an integral language of nightscape lighting, a warm-town concept is delivered. Through focusing on traditional "Alps" truss roof structure and veranda with its structural support, the whole building is blended into the environment. Through the projection of veranda and flood lighting toward stones surrounding the building, the connection with the entire garden is further consolidated.

16

17

18

16 酒店外立面夜景
17 酒店入口夜景
18 酒店夜景

WANDA IBIS STYLES HOTELS CHANGBAISHAN / WANDA HOLIDAY INN EXPRESS CHANGBAISHAN

长白山万达宜必思·尚品酒店 /长白山万达智选假日酒店

开业时间 2013 / 11 / 08
建设地点 吉林 / 长白山
客房数量 1500 间
建筑面积 10.46 万平方米

OPENED ON NOVEMBER 8 / 2013
LOCATION CHANGBAISHAN / JILIN PROVINCE
GUEST ROOMS 1500
FLOOR AREA 104,600 m^2

OVERVIEW OF HOTEL
酒店概述

该项目位于万达长白山度假区西入口处，占地3.24公顷，建筑面积10.46万平方米，共3栋酒店。其中长白山万达宜必思·尚品酒店是雅高集团在中国区第一家度假型酒店；长白山万达智选假日酒店是假日集团在中国区规模最大的一家酒店。3栋酒店钥匙数共计1500把，可以为宾客提供住宿、美食、会议、健身、赏雪、登山全方位的度假体验。3栋酒店依山就势、蜿蜒排布、头尾相连，形成一个聚气藏宝、别有洞天的围合空间。

酒店立面采用质朴的北欧风格，与长白山遥遥相望，落日的余晖，轻轻洒到它的身上，使之完美地融入长白美景之中，静静地矗立在桦林松海之中。

Located at the west entrance of Wanda Changbaishan Resort, the project, with occupied land area of 3.24 hectares and floor area of 104,600 m^2, boasts of three hotels, among which Wanda Ibis Styles Hotels Changbaishan serves as Accor Hotels' first resort hotel in China and Wanda Holiday Inn Express Changbaishan as Holiday Inn's largest hotel in China. With 1500 keys, the three hotels render guests a full holiday experience ranging from accommodation, fine food, meeting, fitness and ski. By depending on the potential shape, winding arrangement and continuous integration, three hotels form an enclosed geomantic treasure land full of discoveries.

Standing apart facing Changbaishan, the hotel facade, with the plain Nordic style, is perfectly blended into the Changbaishan beauty by towering quietly among the sea of birch and pine trees as the sunset tenderly sheds glow on it.

1 酒店总平面
2 酒店鸟瞰

3

3 酒店群
4 酒店外立面

EXTERIOR OF HOTEL
酒店外装

和谐共生，又另辟蹊径——该项目采用质朴的北欧风格，为整个度假区的酒店区奏上和谐的音符，又在松林云海中留下淡淡的痕迹。钟楼、老虎窗、褐红与象牙白色交相辉映，在高低起伏的形体变化中体会到别样的度假感受。

把粗犷的北欧风格作为这个项目的基调，外装设计既与一期酒店风格协调，又能反映出自己的特质，也能符合长白山的雪文化。酒店空调系统的改变让立面多了几排室外空调机位，当煞费苦心将室外空调机位布置好后，空调系统又改回到原来的形式。激动人心的结局总是向着好的一面发展，而3栋楼形体上高高低低的变化又为立面带来了妙趣横生的感觉。

Harmonious coexistence while innovative pursuit-the project adopts plain Nordic style, playing harmonious notes for the resort hotel as a whole and leaving faint traces in the sea of pine trees and clouds. Bell tower, dormant window, and startling interplay of brown red and ivory white deliver a distinctive holiday feeling in the undulating shape-shifting.

Defining rough Nordic style as the keynote for the project, the exterior design harmonizes with the first phase hotel style, reflects its own features, and accords with the snow culture of Changbaishan. The change of hotel HVAC once added several rows of outdoor air conditioners in the façade, but the HAVC is soon restored upon conscientious placement, which denotes that stirring outcome invariably marches toward bright side. The jagged shape of the three hotels, however, brings a sense of fun for the façade.

4

塔楼位于万达长白山智选假日1号酒店的端部，正对园区的主入口，作为该项目的精神堡垒，形体和细节的设计都是围绕这个重点来进行的。从夸张的木构形式，到繁冗的哥特风格，再到精致的Art Deco，经过不断的推敲，最后选择了朴拙的北欧形式，在近乎原始的山林里，用质朴的建筑语言来表达对这方土地的热爱。

Lying at the end of Wanda Holiday Inn Express Changbaishan No.1 and directly facing the main entrance of the park, the tower building, as the spiritual fortress of the project, is the focus in terms of shape and detail design. Through continuous scrutiny over exaggerated wooden form, burdensome Gothic style and exquisite Art Deco, plain Nordic style becomes the winner, as it, in the midst of nearly original forests, expresses its ardent love for this land with a plain architectural language.

5

5 塔楼效果图
6 塔楼实景图

6

7

7 智选假日酒店入口雨棚
8 宜必思·尚品酒店入口雨棚

在雨棚的设计中，首先通过风格迥异的造型来将3栋酒店区别开来；其次将雨篷底边距地的高度不低于5.1米，目的是满足大巴车的通行。完工后的雨棚恢宏大气，漂亮异常。

With regard to the awning design, firstly distinctive awning shapes distinguish the three hotels, and secondly the no less than 5.1-meter height of awning bottom to the floor satisfies the access of bus. All these efforts open up a magnificent and stunning awning.

8

NIGHTSCAPE OF HOTEL
酒店夜景

该项目夜景的立意是“雪原上的珍宝”，通过在建筑基座、腰线及上部檐口运用洗墙灯，把建筑由下而上，由暗渐亮，分层次有重点地洗亮，将这个立意作了完美的诠释。尤其值得一提的是钟塔的照明设计，作为建筑群核心的塔多是以远光投射到塔上以打出辉煌的效果，但是长白山地区冬季积雪非常厚，在屋顶上布置远光灯，若是做低了会被雪埋住，若是做高了，又违背“见光不见灯”原则。经过沟通和模拟设计，最终确定在塔的檐口上做4盏平行坡屋面的洗墙灯，在塔尖处放置4盏射灯，洗墙灯虽然覆盖在冰雪之下，但是透过冰雪的灯光会更璀璨，产生更意想不到的效果。

Adhering to the theme of "Treasure on Snowfield", flood light is adopted at building base, belt course and upper cornice to shine the building from bottom to top and from dark to bright in a hierarchical and focused manner, perfectly interpreting the theme. It is particularly worth mentioning the lighting design of clock wall. Conventionally, the brilliant effect of tower (the core of architectural complex) is created by distance light projection, but Changbaishan is an exception. Since the snow here in winter is extraordinarily thick, the distance light on the roof will be buried by snow in lower position and will violate the principle of "Visible Ray While Invisible Light" in higher position. Through negotiation and simulation design, application of four flood lights parallel with pitched roof at tower cornice and four spotlights at tower tip is finalised. In this case, flood lights, though covered by the snow, will be brighter by penetrating the snow, thus exerting unexpected effect.

9 智选假日1#酒店夜景
10 智选假日2#酒店夜景
宜必思·尚品酒店夜景

9

10a

10b

11 酒店夜景鸟瞰图

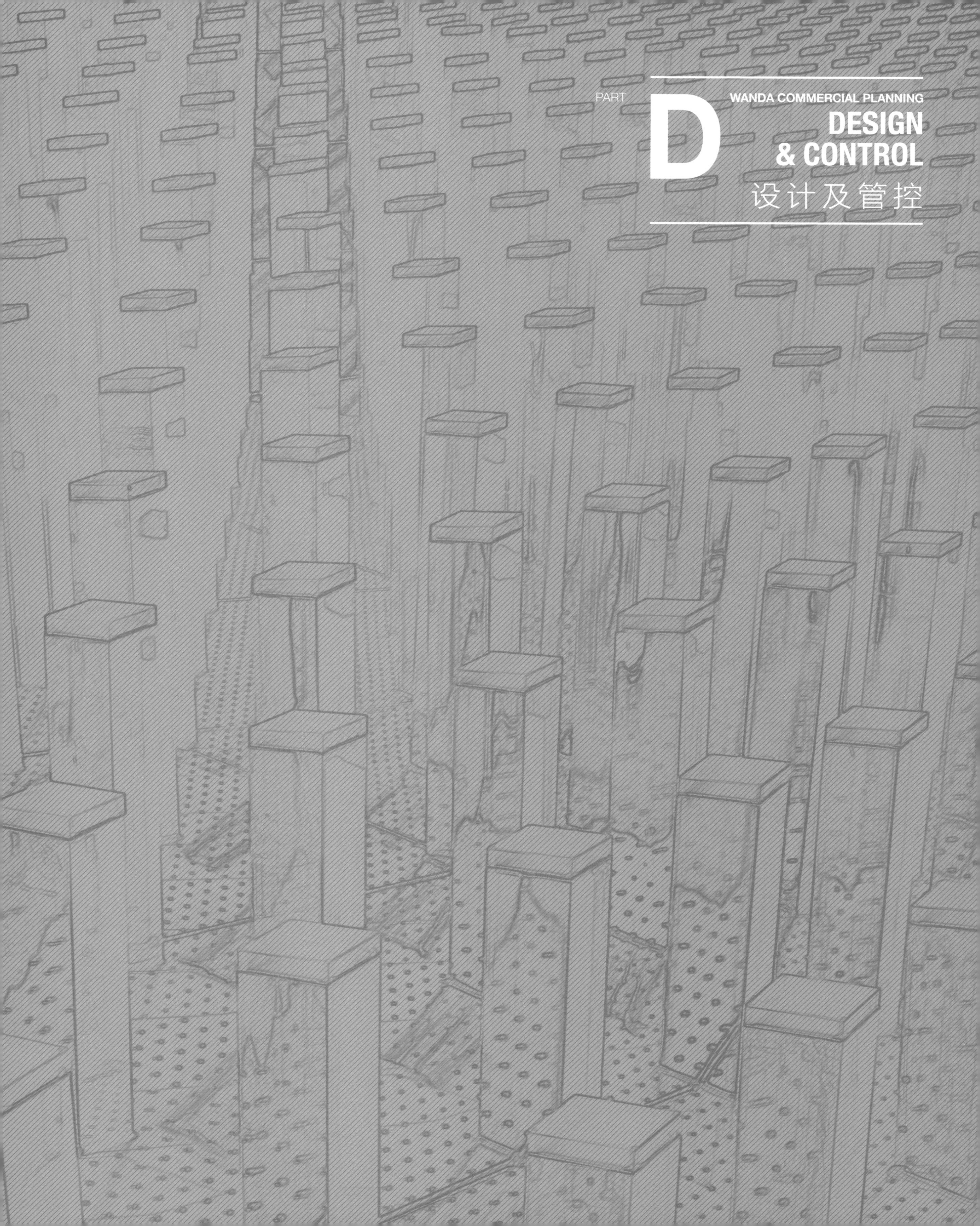

PART

D

WANDA COMMERCIAL PLANNING

DESIGN & CONTROL

设计及管控

OVERVIEW OF PLANNING DESIGN FOR WANDA EXTERIOR PEDESTRIAN STREET

万达室外步行街规划设计概述

文 / 万达商业规划研究院副院长 朱其玮

万达室外步行街是以步行交通为主的商业街，兼具休闲、娱乐等功能。它是万达广场重要的组成部分，它的规划与建设，对于塑造城市区域形象，丰富商业业态，提升城市商业品质，都有着非常重要的作用。万达室外步行街的规划设计，从研究商业特性和经营功能出发，结合当地的历史传统、文化特色，以人为本，创造出商业配置合理，空间形式丰富，环境优美舒适，交通便捷，并富于文化特色的街区环境，以满足人们的物质和精神的双重需求。

万达室外步行街的规划设计，分总体规划、建筑设计、景观设计、夜景设计、导视设计、机电设计和结构设计共7个方面。本文仅就总体规划、建筑设计和景观设计三个方面作简要阐述。

一、总体规划

室外步行街总体规划应着重处理好以下几方面问题。

1 选址要求

万达室外步行街选址一般取决于万达购物中心，从借用购物中心的客流辐射影响力的目的出发，通常毗邻万达购物中心建造，依托购物中心向外辐射，与万达城市综合体住宅底商共同形成了商铺群，并与购物中心等业态，共同组成了业态互补、互相借势、良性互动的商业建筑群。

2 规划原则

区别于其他商业建筑，步行街的规划重点有两方面要求。

1) 步行街应面向城市主要道路开口：作为客流拉动的关键，口部是最重要的形象特征点，须朝向城市主要道路，形成步行街形象展示与客流引力场（图1）。

2) 应避免街包店："街包店"型步行街，外部形象相对封闭，行走动线不合理，与城市环境互动不足，消费客流引导不利，对于后期商业经营弊端较多，应予避免（图2）。

3 步行街动线

街道走向应因地制宜，根据周边道路情况、客流走向、业态分布和地形条件等确定不同动线形式。同时，动线上的交叉路口、街角转折处应适当设置扩大

Wanda Exterior Pedestrian Street, mainly for pedestrian traffic, with supported recreation and entertainment features, is an integral part of Wanda Plaza, thus it's planning & construction is of great meaning to build urban area image, enrich commercial format and escalate commercial quality. Through the study of commercial features and operation functions, and integration of local historical traditions and cultural traits, the planning design of Wanda Exterior Pedestrian Street, holding high of people-oriented ethic, forges a block environment rational in commercial configuration. It is rich in space form, comfortable in environment, convenient in transportation and abundant in cultural features, thus catering to cater for individual's demands spiritually and materially.

The planning design for Wanda Exterior Pedestrian Street consists of master plan, architectural design, landscape design, nightscape design, sign design, mechatronical design and structural design, while only the first three aspects are addressed briefly here.

I. MASTER PLAN

The master plan of exterior pedestrian street shall focus on the issues below.

1 SITE REQUIREMENTS

Generally, Wanda shopping center speaks for the site of Wanda Exterior Pedestrian Street. In order to capitalize on the radiation influence of customer flow, the pedestrian street is typically adjacent to Wanda

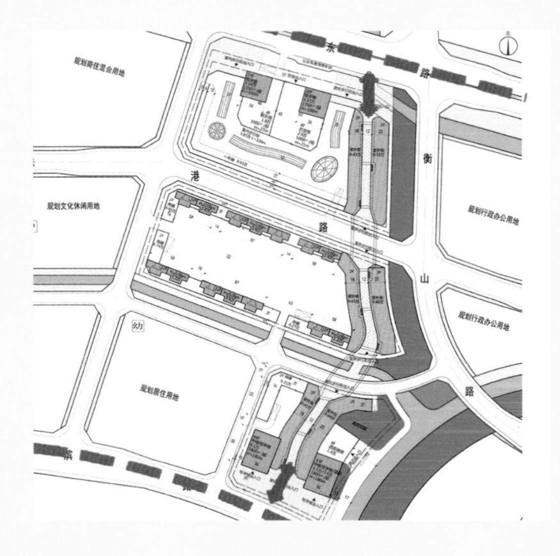

（图1）步行街开口示意

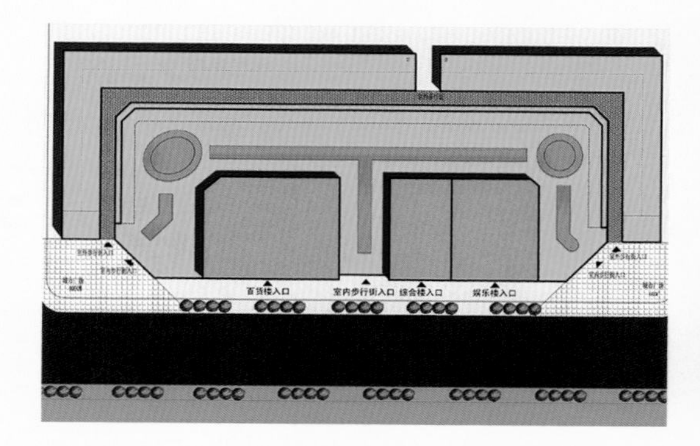

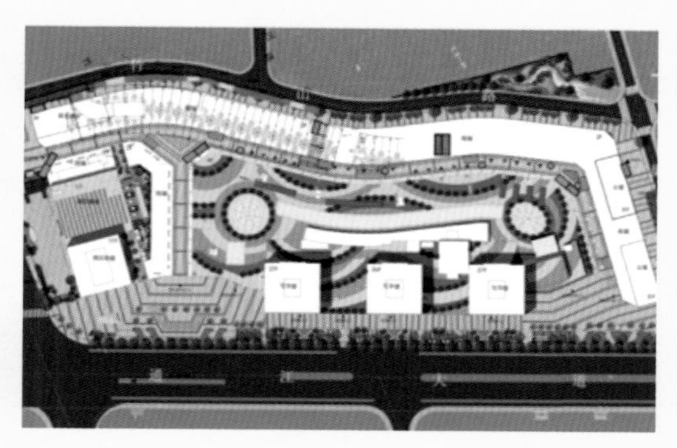

(图2) "街包店"

的节点广场，作为客流转换枢纽及空间环境焦点，从而尽量延长顾客停留时间。

具体动线规划可分"一"字型、"T"字型、"十"字型、"L"型和不规则型等多种类型（图3、图4）。需要说明的是，"一"字型街由于空间形态单调，容易形成"一览无余"而使得逛街体验较差，故应慎重设计。

4 面积规模

根据城市级别、所处区位、半径5公里范围内人口数量、当地商业资源、消费习惯、气候等因素，综合确定适合的步行街商铺规模。

5 技术要求

步行街的规划设计，有多个方面的技术要求，主要有以下3方面。

1)室外步行街宽度:一般控制在12~15米，单侧临河的酒吧街，街道宽度可加宽至18~25米。

2)街铺建筑高度控制：以2~3层为宜，商铺高度与街道宽度比值宜控制在1：1。

shopping center, hoping to form the boutique shop cluster together with residential floor traders of Wanda urban building complex by radiating outward in virtue of shopping center, and the commercial building cluster (complementary in format, dependent on each other's power and benign in interaction) with commercial format like shopping center.

2 PLANNING PRINCIPLES

Unlike other commercial buildings, planning of the pedestrian street mainly covers two aspects.

1) The pedestrian street shall be geared to the urban main road: as gateway, the key of customer flow drive, serves as the most prominent image feature point, the pedestrian street shall be geared to the urban main road, forging street image display and customer flow attraction field (Fig. 1).

2) "Street Encircling Shops" shall be avoided: such type of pedestrian street shall be avoided because it is somewhat adverse to the following commercial operation due to its relatively closed external image, irrational circulation, inadequate interaction with urban environment and improper guidance for consumers (Fig. 2).

3 CIRCULATION OF PEDESTRIAN STREET

Following the ethics of location, circulation forms vary depending on the surrounding road conditions, customer flow trend, commerce distribution and terrain conditions. Meanwhile, at the intersections and street turns of the circulation, enlarged plaza nodes shall be established to serve as the flow switching hub and spatial environment focus, thus extending customers' stopover utmost.

The specific circulation can be planned as line-type, T-type, cross-type, L-type and irregular types (Fig. 3 and Fig. 4). It is noted that special attention shall be paid to the line-type street, since the panoramic view brought by its monotonous spatial form will give rise to an undesirable shopping experience.

4 SCALE OF AREA

Determine appropriate scale of pedestrian street shops as per the city level, location, population within a 5km-radius, local business resources, consumption habits and climate.

5 TECHNICAL REQUIREMENTS

The planning design concerned involves technical requirements on several aspects, mainly on three ones below.

1) Width of exterior street: typically within 12~15 meters, and 18~25 meters for bar street with either side adjacent to river.

2) Height control of street shops: 2~3 storeys are preferred, and the ratio between shop height and street width shall be limited to 1:1.

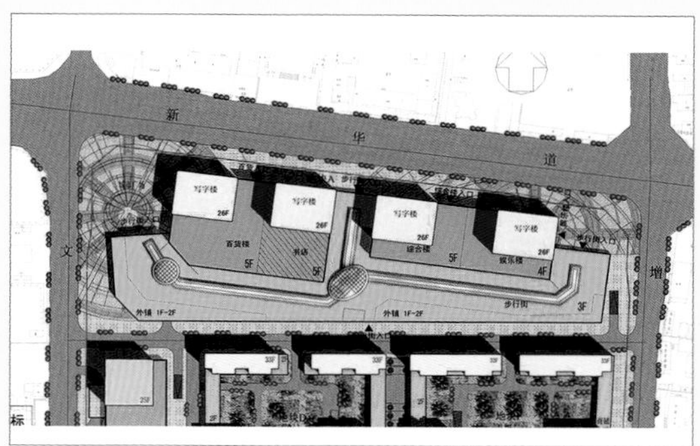

"一"字型

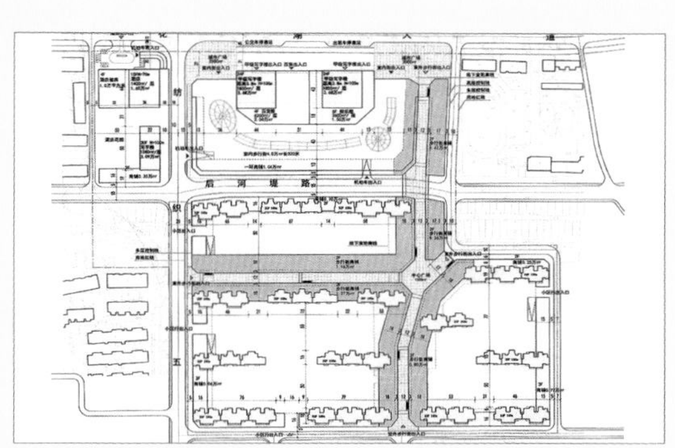

"T"字型

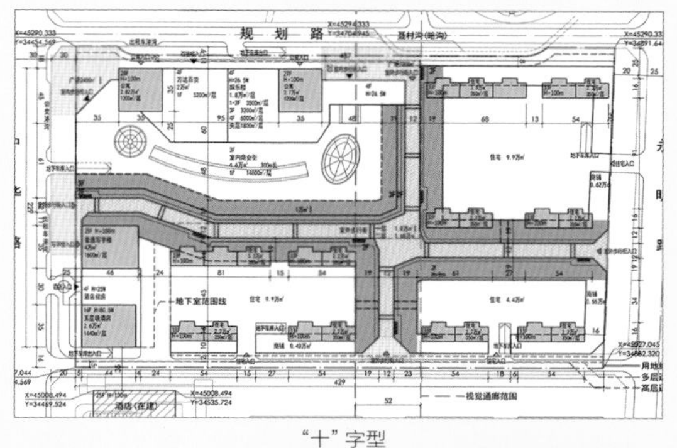

"十"字型

(图3) 步行街动线类型一

3) 人车分流：合理规划车流人流路线，做到人车分流；对于车库出入口的设置，不能影响步行街人流通行及商铺经营。

二、建筑设计

建筑设计分平面设计、立面设计、节能设计三个方面。

1 平面设计

室外步行街一般为两层建筑，分跃层式商铺及分层销售式商铺。单铺面积控制在120~140平方米。单铺的开间一般为4~5米，进深10~20米。

室外步行街应按比例设置一定的辅助用房和设施，包括公共卫生间、货梯、垃圾房、物管用房和电表间等。平面设计应考虑后期商铺合并的需求，结合结构柱网设置楼梯间、消火栓、空调位、管井、烟道及商铺内机电点位等功能性配套设施。

2 立面设计

立面设计的宗旨是，符合城市规划以及商业综合体对商铺立面效果的总体要求，结合地域文化特色，并满足“一店一色”、“经营展示”等商业要求的造型丰富、优美的外立面效果。立面设计重点关注以下几个方面：

1)重点部位设计。主入口、节点小广场、内外街联通口为立面设计的重点部位，应强化处理，加强标识性及商业氛围。

2)层高。商铺一般首层层高4.5~5.7米，二层层高4.2~5.1米。

3)商业展示。立面设计应在满足节能要求的情况下最大程度考虑商业展示的需要，尽量使得经营动态可对外展示。

4)立面造型及材质。建筑平面宜适当凸凹，形体宜错落有致，天际线应尽量丰富多变，避免平直的单调效果；其次，立面造型、构图宜采用模块组合变化，形成丰富多变、一店一色的效果。

3 节能设计

建筑节能是设计中重要的环节，万达建立了自己的节能指标体系，节能效果逐年提升。

三、景观设计

基于消费者对商业环境日益增长的要求，及希望更加生活化的购物方式和场景化的休闲氛围等现实因素，室外步行街的景观设计应遵循主题、功能、场景和安全等原则来进行。

3) Person-vehicle divergence: realize person-vehicle divergence by planning the traffic and customer flow routes in a rational way. Access of garage shall be set in such a way that it will not affect the passage of customers on street and shop operation.

II. ARCHITECTURAL DESIGN

Architectural design is composed of graphic design, façade design and energy-saving design.

1 GRAPHIC DESIGN

Generally with two-storey buildings, the exterior pedestrian street has duplex apartment shops and layered sale shops, each being 120~140 m^2 in area, 4~5 meters in bay and 10~20 meters in depth.

Exterior street shall be furnished with some auxiliary buildings and facilities by proportion, including public washroom, freight elevator, garbage room, property management room and meter room. In view of the shop merger demand in later period, graphic design should, combing with structural column grid, set up functional facilities such as stairways, fire hydrants, air-conditioned spaces, tube wells, flues and electromechanical points in shops.

2 FACADE DESIGN

Adhering to the philosophy of conforming to the general requirements on shop façade effect by urban planning and urban building complex, incorporating regional culture characteristics and catering for an enriched and graceful façade effect in line with the commercial requirements on "One store, One Style" and "Operation Exhibition", the façade design focuses on the aspects below:

1) Design of key sites: main entrance, node plazas, and linking access of exterior & interior streets, as the key sites of façade design, shall be highlighted by enhancing signage and commercial atmosphere.

2) Storey height: as a rule, the 1st Floor of shops shall be 4.5~5.7 meters high and 2nd Floor 4.2~5.1 meters high.

3) Commercial exhibition: where energy-saving requirement is satisfied, the façade design shall, to the largest extent, consider the need of commercial exhibition by displaying operation dynamic utmost.

4) Façade shape and materials: it shall be somewhat bumpy in architectural plan, well-proportioned in form, rich and varied in skyline to avoid a monotonous effect of being only straight; variation of combined modules shall be adopted in façade shape and mapping to assume the varied and one shop one feature effect.

3 ENERGY-SAVING DESIGN

Perceiving building energy efficiency is an important link in design. Wanda has established its own energy-saving indicator system, harvesting yearly increasing energy-saving effect.

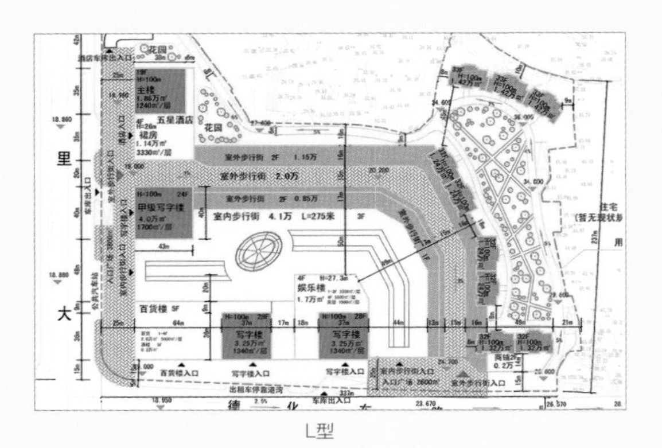
L型

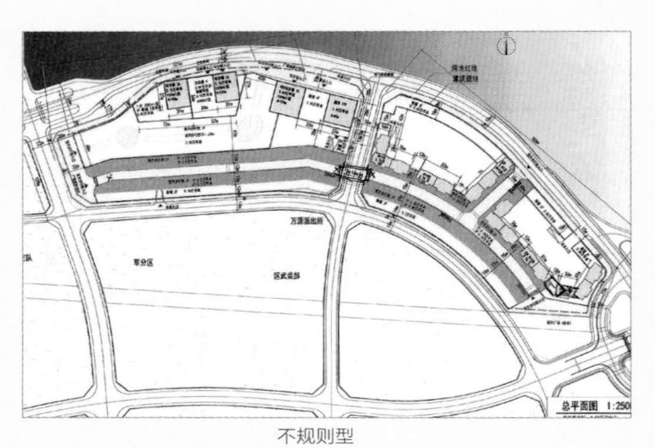
不规则型

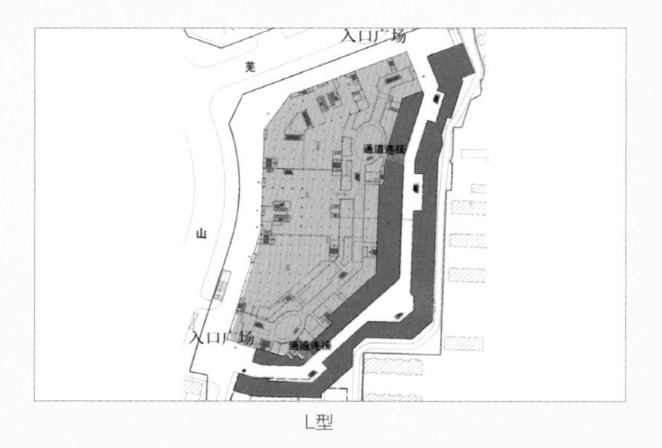

L型

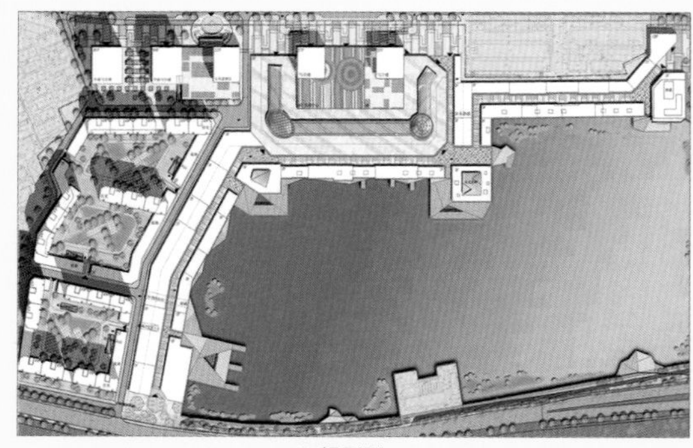
不规则型

(图4) 步行街动线类型二

1 主题原则

室外步行街景观设计需遵循一致性、地域性和故事性的主题设计原则。景观应与建筑特色保持一致，并深度挖掘地域文化特色，利用所有景观元素体现主题文化，并使故事主题沿同一主线贯穿于整条室外步行街中。

2 功能原则

室外步行街景观需符合环境基本功能和商业功能，并利用不同功能类型的景观元素，加强商业环境的引导、停留、休憩、娱乐和无障碍等服务功能。

3 场景原则

在室外步行街景观环境营造中起到至关重要作用的是场景原则，它是室外生活场景的模拟再现，也是吸引消费者停留休憩的关键环节。根据室外步行街的商铺业态分布、空间特点和人行为特点等，将所有景观元素进行混搭组合，进而展现每条步行街的商业生活特质。

4 安全原则

室外步行街环境须满足消防安全和人身安全的要求。消防方面，应注意不能对场地消防车道、消防扑救面形成障碍。人身安全方面，首先是材料选择上应注意使用环保无毒材料、无刺无毒植物；其次，造型方面应注意在人可触摸范围内不应有尖锐棱角等。

综上所述，万达室外步行街的规划设计必须遵循商业运行的规律，遵循自身的发展与建设规律，以此为基础，确定相应的规划设计要求。在未来几年内，室外步行街将向综合化方向发展，由单纯购物功能扩展至购物、游览、休闲等多种功能；由主要进行街景等外部硬件的构建，调整为人文内涵、人性化理念的营造。这也是在万达室外步行街的规划设计领域，时代给我们提出的新课题。

III. LANDSCAPE DESIGN

In light of consumers' ever-growing demands on commercial environment, and their expectation on ever life-oriented shopping mode and scenarized recreaion atmosphere, the landscape design of exterior pedestrian street shall be implemented in line with the theme, function, scene and safety principles.

1 THEME PRINCIPLE

Landscape design of exterior street shall follow the theme design principle of consistency, regionalism and storytelling: Landscape shall be consistent with architectural features, regional culture features shall be excavated profoundly, all landscape elements shall be utilized to embody the theme culture, and the story theme shall be around the same plotline to go through the whole street.

2 FUNCTION PRINCIPLE

Consistent with the basic function of environment and commercial function, the landscape of exterior street shall, through utilizing the landscape elements of various function types, intensify the service function like guidance, stopover, recreation, entertainment and accessibility in commercial environment.

3 SCENE PRINCIPLE

Being the simulation and reconstruction of exterior scenes of life, and the key link of attracting consumers for stopover and recreation, scene principle plays a crucial role in creating the landscape environment of exterior street. Mixed combination is applied to all landscape elements as per commercial distribution of boutique shops, spatial features and consumers' behavior characteristics of exterior street, hoping to exhibit commercial life qualities of each pedestrian street.

4 SAFETY PRINCIPLES

Exterior street environment shall be fire safety and personal safety satisfied. As for the former one, site fire lane and fire-control platform shall be unobstructed. For the latter one, nontoxic materials and plants without thorns are desirable in terms of material selection, and sharp edges within the range of human touch shall be unavailable in terms of shape.

To sum up, the planning design of Wanda Exterior Pedestrian Commercial Street must follow the laws of commercial operation and its own development and construction, and based on which, the corresponding planning design requirements are determined. For years to come, Exterior Pedestrian Street tends to evolve toward integration, reaching out to multiple functions like shopping, sightseeing and recreation beyond its pure shopping function, and adjusting to the creation of humanistic connotation and concept of humanity from the original construction of external hardware like streetscape. This is the new subject brought by era in the planning design field of Wanda Exterior Pedestrian Street.

INTERIOR DESIGN AND CONTROL OF WANDA PLAZA: OVERALL CONTROL & PRECISION MANAGEMENT

万达广场内装设计与管控——全程管控，精细管理

文 / 万达商业规划研究院室内装饰所副所长 张振宇

在万达广场的开发建设过程中，内装的设计与管控是项目整体设计管控中极为重要的一环。内装专业的设计与管控不仅仅是传统意义上对设计单位的管理，而是从概念规划、方案设计、施工图设计、材料封样、现场实施阶段管控到运营整改、复盘总结、成果转换等全方位的过程管控（图1）。

按照内装专业设计管控内容，分为图纸管控、设计生产力管控、材料管控、现场实施阶段管控、资料管理及成果转换管控等五个方面。

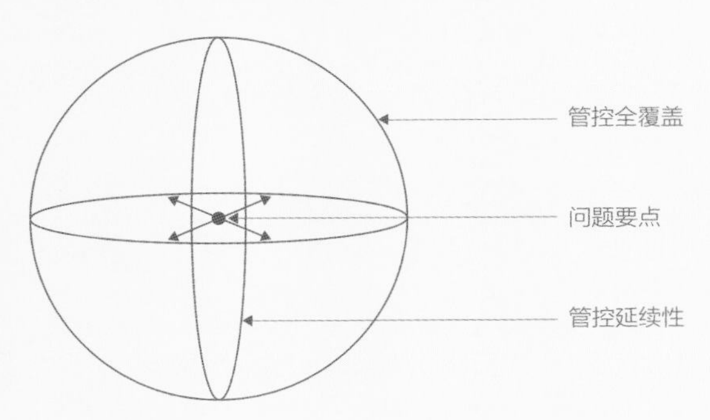

（图1）内装专业管控思维模型图

一、图纸管控

内装专业设计管控分为：阶段管控、重点管控和专项管控三个方面。

1 阶段管控

阶段管控过程，包括步行街内装的设计方案招投标管控阶段、深化设计管控阶段、一版施工图管控阶段、二版施工图管控等四个阶段。在施工图管控阶段，应结合内装定标方案、建筑及各专业图纸、二次机电配合等方面综合进行，避免一、二版施工图纸产生较大修改，同时结合模块计划严控出图时间，确保招标、施工等后续计划节点按时完成。

2 重点管控

依据已签字确认的方案图纸，指导施工图深化设计。注意与签字确认的内装方案风格不应产生较大偏差，保证原定方案风格这条主“线”，同时要使方案效果可以在图纸上建立并最终确保实现。努力将签字确认方案这条主线作为工作的思想纲要贯穿在设计、实施等各环节中。

In the process of development and construction of Wanda Plaza, design and control of interior, as the extremely significant part of project overall design and control, is not merely the management on design units in the traditional sense, but the all-around process control involving concept plan, scheme design, construction drawing design, sealed material sample, site implementation phase, operation rectification, checking summary and result conversion (Fig. 1).

As per content, the design and control of interior can be divided into five aspects, including drawing control, design productivity control, material control, control of site implementation phase and control of data management & result conversion.

I. DRAWING CONTROL

Design and control of interior specialty consists of phase control, major control and special control.

1 PHASE CONTROL

The phase control goes through four phases covering design scheme bidding control of pedestrian interior, deepening design and control, Version I construction drawing and Version II construction drawing. The construction drawing control shall be, through integrating successful bidding scheme of interior, architecture and other specialty drawings, secondary electromechanical cooperation, implemented comprehensively to avoid major revision incurred to Version I and II construction drawings, and shall, via combining module plan, strictly control the drawing-completion time to ensure the follow-up plan nodes such as bidding and construction to be completed on schedule.

2 MAJOR CONTROL

Direct the deepening design of construction drawing, which shall not deviate too much from the interior scheme style upon signature verification, as per scheme drawing with signature verification. Ensure the original scheme style serves as the "Main Line", and its effect is established in the drawing and realized ultimately. Endeavor to apply the main line-work philosophy to every link incorporating design and implementation.

3 SPECIAL CONTROL

In the special phases such as interior sample section and full implementation of site construction, interior effect and quality are sure to be achieved ultimately by organically integrating construction drawing and site to make the intent of design drawing remain unchanged

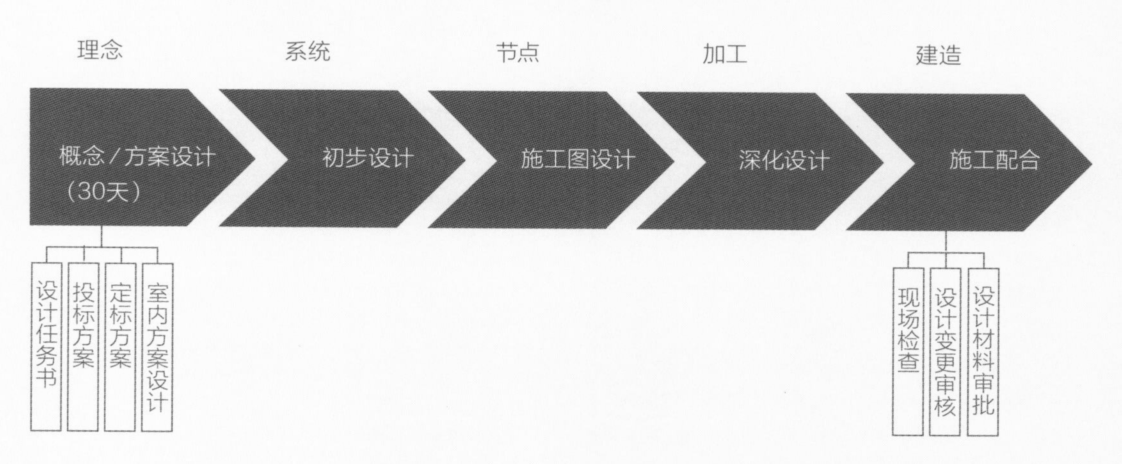

(图2) 内装方案、施工图阶段设计管控流程图

3 专项管控

在内装样板段、现场施工全面实施等专项阶段，要保证内装效果及质量的最终实现，应充分把施工图纸同现场有机结合，确保在现场不改变设计图纸意图，协调土建、机电、结构等各专业协助处理各类现场问题，保证内装设计效果的顺利实现（图2）。

二、设计生产力的管控

内装专业设计人员的技术水平直接影响内装效果、品质，对内装专业各家供方单位的主创人员、施工图人员要遵循先培训后上岗的原则，要将万达集团大商业及内装专业的各项强制条例、设计要求、设计标准、管控要求、配合服务要点等各项规章制度进行专业培训及交底，设计人员熟练了解后方能允许其进行室内街内装设计。具体管控过程如下。

首先，应了解、考核内装专业各家供方单位设计人员整体素质、设计水平、审美修养、服务配合等。对于重点项目、重点部位、重要时间节点等，需要集中优势单位进行应对，以保障项目整体效果及按计划顺利完成。

其次，各项目设计展开后，应督促设计单位进行全面的计划安排，结合计划模块，分解细化整体工作计划，明确工作目标、质量要求、时间节点、技术对接负责人等。从而最大程度避免不必要的时间成本和设计资源的浪费。

on site, and by coordinating civil engineering, electromechanical and structure specialties to help address site problems (Fig. 2).

II. CONTROL OF DESIGN PRODUCTIVITY

As the technical level of interior designers is directly related to the effect and quality of interior, the creative staff and construction drawing personnel from interior supplier units shall cling to the principle of training first, during which all rules and regulations shall be professionally trained and disclosed, involving mandatory regulations, design requirements, design criteria, control requirements, coordinated service points of Wanda Group big business and interior specialty. Designers are only allowed to commence on the interior design of interior street when they are well informed of the training content. The specific control process is as follows.

Firstly, it is necessary to understand and assess the overall quality, design level, aesthetic accomplishment and service coordination of designers from interior supplier units, and to converge preponderant units to address the major projects, key sites and significant time nodes, aiming to guarantee the overall effect of project and completion on schedule.

Secondly, when each project is in full swing, it is necessary to urge the design unit for making overall schedule, breaking down overall work plan via integrating plan module, and identifying work objective, quality requirement, time node and responsible person

三、材料管控

材料管控主要为技术交底、样板确认、样板段实施等三个阶段。

1 技术交底
在现场正式大面积实施前，规划院室内装饰所组织内装设计单位对施工单位进行图纸的设计交底工作。确保施工单位理解设计图纸中的重点与难点。明确样板先行的原则，及时协调设计单位进行各种材料的技术交底，确保品质、效果、成本三要素全面落实。

2 样板确认
装饰材料使用的好坏直接决定内装的效果及安全，所以在材料管控时应重点控制好设计封样及施工材料样板的确认工作，达不到设计样板要求的，坚决不允许进场使用。内装材料的设计确认及把控应纳入项目管控的工作计划中，确保各类材料封样及确认的及时性、品质保证性、材料安全性等，材料封样及确认应根据项目进度计划分批次进行，进行动态线性管控。

3 样板段实施
样板段实施是检验设计效果和减少成本浪费的重要一环。保证材料必须先上样板段实施，以检验材料实施工艺与效果，确保内装效果、品质、安全、计划的全面实施。

四、现场实施阶段管控

内装现场实施阶段管控，主要以中期检查和规划验收为主要手段。中期检查和规划验收工作，对于监控内装实施过程中质量、安全、品质、效果起到关键作用。中期检查、验收，可及时发现问题并及时纠正，避免施工过程中的不良环节形成恶性循环。过程检查、验收环节，以明确问题、纠偏方案、完成时效、反馈上报等为工作要点。

中期检查、规划验收工作不要停留在表面或书面上，更不能停留在某个项目某个阶段上，要把中期检查、规划验收工作当作质量、品质、效果保证的一种管控手段。内装专业中期检查、验收要全面考量，既要检验工程是否按图施工等工程问题，也要检验施工图纸、材料封样等前期设计问题。同时，中期检查、验收环节应保证各参与方全部到场，要做到验收图纸资料齐备，并要引起各部门（尤其是项目公司领导）的高度重视与关注（图3）。

of technical exchange, hoping to, to the largest extent, avoid unnecessary waste on time cost and design resources.

III. MATERIAL CONTROL

Material control mainly consists of technical disclosure, sample verification and implementation of sample section.

1 TECHNICAL DISCLOSURE
Prior to widespread site implementation, Interior Decoration Office of Planning Institute shall organize interior design unit to proceed technical disclosure of drawings to construction unit, ensuring the latter has interpreted the emphasis and difficulty in design drawing, identified the principle of sample first and will coordinate with design unit for technical disclosure of materials in a timely manner to safeguard the full implementation of quality, effect and cost.

2 SAMPLE VERIFICATION
Effect and security of interior is directly dependent on quality of decorative materials, thus during material control, sealed samples and verification of construction material samples shall be under critical control, forbidding the site use for those failing to satisfy the design sample requirements. The design verification and control of interior materials shall be incorporated into the work plan of project control to ensure the promptness, quality assurance and material safety of sealed samples and verification of materials, which shall be implemented by lot as per project schedule and subject to dynamic linear control.

3 IMPLEMENTATION OF SAMPLE SECTION
Implementation of sample section serves as a significant link in testing design effect and eliminating cost, thus materials must be subject to this link, aiming to test the implementation process and effect of materials and guaranteeing the full implementation of effect, quality, safety and plan of interior.

IV. CONTROL OF SITE IMPLEMENTATION PHASE

The control of interior site implementation phase mainly relies on intermediate inspection and planning acceptance, which play crucial role in supervising quality, safety, taste and effect in the process of interior implementation. Intermediate inspection and planning acceptance, mainly work on problem identification, scheme rectification, completion prescription, feedback report, are capable of timely spotting and addressing problems to avoid vicious circle incurred by undesirable link during construction.

Intermediate inspection and planning acceptance shall not be still in superficial or paper form and even worse, target at specific project or specific phase. Rather, they

五、资料管理及成果转换管控

资料管理工作在设计管控中起着重要的作用。考核工作完成情况，回顾项目管控的各项事件等，都需要第一手资料完备。因此需要把资料管理工作从"点"抓起，点滴工作细节汇总、文件的随时整理，如方案设计中的一张草图、一块推敲的材料样板、一份管理过程中的签字文件等，点滴累计收集，就能形成项目的整体设计档案。内装资料管理工作应做到按期汇总、重要资料应按照项目节点计划整理上报，所内应按月审核，对没有及时提交资料的人员应及时提醒，以保证资料管理这条线无疏漏，做到及时、全面、完整。

资料管理工作应做到对专业性较强的资料及时进行科研成果转换，做到技术经验共享并进行全面推广。例如《内装定额标准图集》既是整理汇总万达广场内装专业重点、难点做法并及时进行科研成果转换，以指导后续项目。成果完成后应组织对相关人员、设计供方进行交底、全员培训。

内装专业设计管控作为贯穿在整个项目过程的管理动作，其影响力之大、覆盖面之广，是有效把控项目最终效果、有效控制项目成本、保证项目按计划实施的重要手段，全面管控、精细管理既是内装设计与管控的精髓也是保证万达广场内装高品质完美实现必要保障。

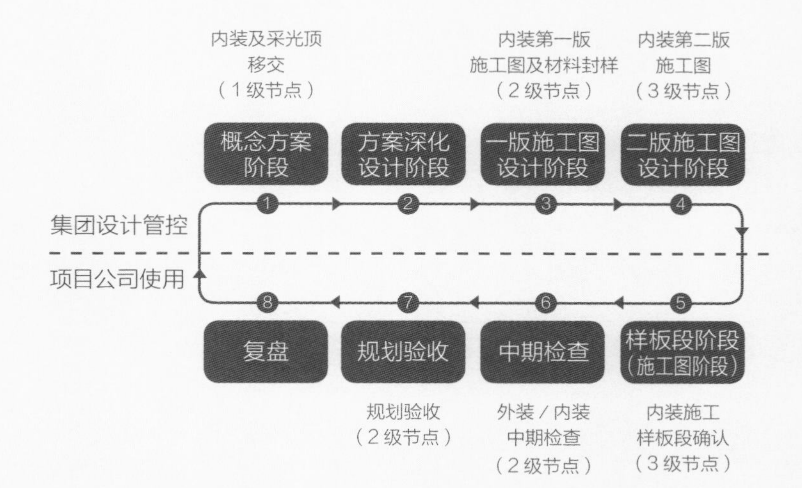

（图3） 设计、实施、运营阶段设计管控流程图

shall be treated as one of the control means aiming at quality, taste and effect assurance. When it turns to interior, such work shall be subject to full consideration. It shall cover the engineering problem-whether the project is constructed as per drawing, and preliminary design issues-construction drawing and sealed samples of materials. Meanwhile, the work shall ensure that all parties concerned shall be present, acceptance drawings shall be complete and that all departments, project company leadership in particular, shall show grave concern to the work (Fig. 3).

V. CONTROL OF DATA MANAGEMENT & RESULT CONVERSION

Data management plays a crucial role in design control in that complete first-hand data is indispensable to assess work performance and to review events in project control. To this end, data management work shall start from "Bit". With detail summary of a bit of work and filing of documents whenever necessary, such as a draft in scheme design, one material sample under scrutiny and a signed document in the process of management, overall design file of project tends to be presented. In terms of interior data management, summary on schedule shall be made; important information shall be sorted and reported as per project node plan; monthly review in office shall be available; the staff failing to submit data timely shall be reminded. In this way, it ensure exempting data management from any omission, making it timely, comprehensive and complete.

Data management shall work on highly specialized data for timely scientific achievements conversion, enabling technical experience sharing and widespread promotion. For instance, by implementing timely scientific achievements conversion for *Atlas of Interior Quota Standard* (a summary of emphasis and difficulty practice in interior specialty of Wanda Plaza), it gives guidance to the follow-up project. Upon conversion, disclosure to relevant personnel and design supplier, and crew training shall be carried out.

In view of its enormous influence and extensive coverage, design and control for interior specialty-the management action running through the whole project, serves as a major means of effectively controlling the ultimate effect of project, the project cost and of ensuring project implementation as planned. While whole course control and precision management, as the essence of interior design and control, safeguard the perfect presentation of high quality interior of Wanda Plaza as well.

LANDSCAPE DESIGN AND CONTROL OF WANDA PLAZA
万达广场景观设计与管控

文 / 万达商业规划研究院景观所所长 高振江

万达广场是由大商业、酒店、商业街（含室外商业街及步行街）、写字楼和豪宅等组成的城市综合功能区。万达广场景观是由以上单项景观及周边辐射区域景观组成。万达广场景观规划体系的建立和设计实现是万达商业规划研究院景观所对万达广场景观管控的主要内容。在设计实现过程中，计划模块化系统是确保进度的重要保证，而设计和管控的不断创新一直是景观管控中追求的目标。

Wanda Plaza is an urban mixed-use area comprising large commercial area, hotel, commercial street (including exterior commercial street and pedestrian street), office building and mansion. Its landscape consists of the single landscape above and landscape in surrounding radiation area. The establishment and design implementation of landscape planning system of Wanda plaza construct the main content in landscape control of Wanda Plaza required by Landscape Office of Wanda Commercial Planning & Research Institute. In the process of design implementation, planning modular system is to ensure the schedule, while continuous innovation is the constant pursuit of landscape design and control.

一、万达广场景观规划体系的功能及特点

万达广场景观规划体系，必须具备城市功能、区域功能和单体功能。城市功能是指商业综合体与城市空间的关系，包括与城市道路、轴线等关系；区域功能是指万达广场区域内各个接驳口之间的关系；而单体功能则更倾向于各构成要素内部的个性特点。而这些功能的实现，需要通过导视系统、道路体系、绿化体系、夜景照明体系、水景体系、铺装体系和雕塑小品体系等景观元素组合。

万达广场景观规划体系首先是由商业功能定位来确定的，同时也是城市空间与商业公共空间景观设计的导则，为景观品质的不断提升提供了基础。这是商业综合体景观，不同于住宅景观、购物中心景观、酒店景观、道路景观等单一类型景观的最大特点（图1）。

I. FUNCTIONS & FEATURES OF LANDSCAPE PLANNING SYSTEM OF WANDA PLAZA

Landscape planning system of Wanda Plaza shall boast of urban function, regional function and single function. Urban function refers to the relation between commercial complex and urban space, including urban roads and axes; regional function refers to the relation between joints in Wanda plaza area; single function tends to denote the internal personality of each constituent element. The realization of these functions asks for combination of landscape elements, including guide system, road system, green system, nightscape lighting system, waterscape system, paving system and sculpture accessories system.

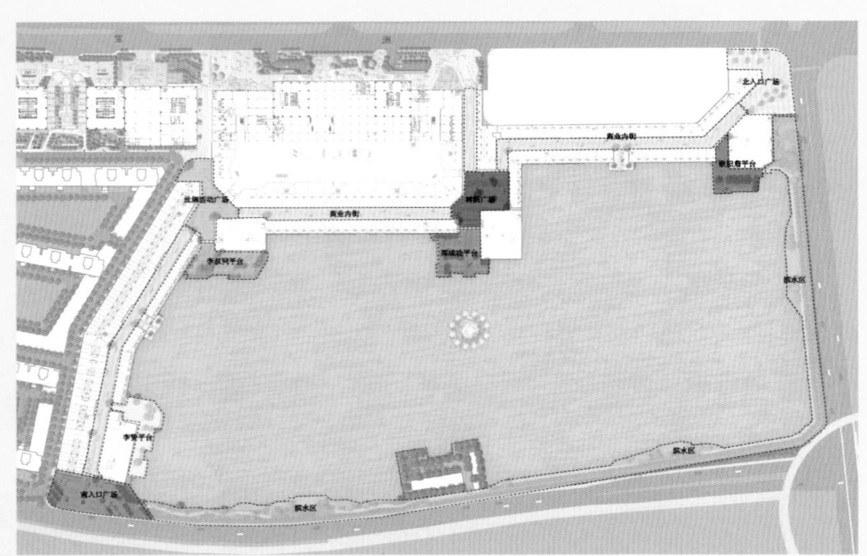

（图1）泉州浦西万达广场

二、万达广场景观管控特点

万达广场景观规划体系是对建筑规划体系的补充和完善，要求具备较强的专业能力、综合协调能力和现场管理能力。计划模块系统是确保万达速度的利器。景观计划模块重点体现在配合前期销售、景观工程招标和景观体系实现三个阶段。在管控的各个阶段对景观体系的建立都有不同的要求。配合前期销售的景观模型方案阶段，确定景观主要功能、主题和风格走向；景观一版施工图是功能、主题和风格落地并将付诸实施的阶段；而根据现场条件协调、管理则是景观二版施工图指导现场施工的主要目的。在实施阶段，景观现场的管控则要求根据实际空间尺度和周围建筑关系，现场确定景观元素的组合关系和比例尺度、材料、肌理等景观细节；同时也是对景观的再设计。为保证计划模块填报的按时保质完成，景观所通过OA限时审批、问题限时处理、模块节点完成前置以及加强与多部门的沟通等方式，为各项目的顺利推进和高品质完成提供了有力保障。

三、景观设计与管控创新

万达广场景观设计创新始终紧紧围绕两点进行：创意和表达。好的创意是基于对当地文化和建筑语言的深刻理解和对商业活动的需求理解之上的。景观主题的选择立足于一个相对具象的切入点，而且还要能耳熟能详，引起当地人共鸣，这样才能准确地抓住融入点。好的景观主题还必须有艺术的表达方法，形成视觉上的美感，形式空间上的舒适放松感。

主题之于景观就像灵魂之于人体。主题是景观贯穿始终的主线，没有主题的景观就是物品的堆砌，杂乱无章。文化则是主题的地域属性，有文化主题的景观是对当地人文的尊重，同时也赋予了强大的文化张力和内涵。没有文化的主题就失去了维持持久生命力的源泉（图2、图3）。

城市商业景观空间是城市公共空间的组成部分，可达性和视线畅通性是必须满足的功能。在视觉和空间结构上，通往商业广场的任何方向应该是通透的。在商业景观空间内部，各个空间要有机组合在一起，在提供视线引导的同时，形成内向的围合空间。这样

Landscape planning system of Wanda Plaza is primarily determined by business functional orientation, and meanwhile, the guideline for landscape design of urban space and commercial public space, which lays foundation for constant enhancement of landscape quality. That is the most prominent feature distinguishes commercial complex landscape from single type landscape such as residence landscape, shopping center landscape, hotel landscape and road landscape (Fig. 1).

II. LANDSCAPE CONTROL FEATURES OF WANDA PLAZA

Landscape planning system of Wanda Plaza, as supplement and improvement for architectural planning system, shall be equipped with strong professional competence, comprehensive coordination capacity and site management capability. Planning modular system serves as a tool to ensure the speed of Wanda, and in terms of landscape planning module, it mainly functions on coordinating early sales, landscape engineering bidding and implementation of landscape system. The requirements on establishment of landscape system vary at each phase of control: in the landscape model scheme phase coordinating early sales, major function, theme and style of landscape shall be determined; in Version I construction drawing phase, the three aspects mentioned above shall be achieved and implemented; in Version II construction drawing phase, coordinate and manage as per site conditions constitute the main purpose of site construction guidance. When it enters into implementation phase, the control of landscape site demands to identify combination relation of landscape elements and landscape details involving ratio scale, material and texture, which is also the redesign for landscape. To ensure the completion of planning module report on schedule with quality guaranteed, the Landscape Office, through OA timed approval, timed treatment for problems, preposed modular node and further negotiation with multiple departments, effectively safeguard the smooth proceeding and high-quality accomplishment of project.

III. INNOVATION OF LANDSCAPE DESIGN AND CONTROL

Landscape design innovation of Wanda plaza on two points throughout: creativity and expression. Desirable creativity is based on a profound perception of local

（图2）东莞长安万达广场

（图3）荆州万达广场

才能加强商业空间的场所感，让人能够停留下来，为商业活动创造条件。在景观空间内部，通过地形的雕塑化，景观小品的造型艺术化和一些趣味性的图案或设施，形成富有美感的景观环境。通过材质、颜色的对比，形成视觉活跃，复杂又和谐的景观环境与体验（图4、图5）。

景观管控创新重点在于解放思想，大胆推陈出新。景观所管控人员在充分理解制度和强条的基础上，结合景观体系和商业功能，不断追求符合商业功能的主题鲜明又富有文化内涵的景观特色，不断寻求不同商业综合体之间的差异性，避免同质化。景观设计供方是实现管控思想的重要环节，通过定期举行供方交流大会，让设计供方充分了解我们的管控思想和要

culture and architectural language and understanding of the demand of commercial activities. Hence, the selection of landscape theme shall be established in a relatively concrete entry point, and better, be popular to arouse sympathy of local people. By virtue of this, the entry point can be spotted exactly. Besides, artistic expressing method is indispensable to a desirable landscape theme in that it facilitates a sense of beauty visually and a sense of relaxation spatially.

Theme is to landscape what soul is to human. As theme is the main line running through the whole landscape, the landscape without theme is merely a pile of disorderly items. While culture serves as the regional attribute of theme: the landscape with cultural theme presents reverence to local humanity and is infused with strong cultural tension and connotation; the theme without culture, instead, loses the source of lasting vitality (Fig.2 and Fig. 3).

Being a part of urban public space, urban commercial landscape space shall cater for accessibility and clear view by ensuring that, in visual and spatial structure, any direction leading to commercial plaza shall be see-through. Inside the commercial landscape space, each space shall be organically integrated to present vision guidance and inward enclosed space, hoping to, through reinforce the sense of place of commercial space to attract customer flow, create conditions for commercial activities. Inside the landscape space, create an aesthetic landscape environment by virtue

（图4）兰州万达广场

（图5）上海松江万达广场

求，以及各自在供方库中的水平层级。为加强内部竞争，景观所还通过建立供方综合信息管理平台和供方全程量化考核体系来不断正向激励创新。

万达广场景观规划体系随着万达业态的丰富，也在不断丰富自己的内涵。目前已经涵盖了购物中心、城市酒店、高档写字楼、旅游度假酒店（含生态环境建立和维护）等各种类型的景观管控，景观所在不断加强内部管理，提升自身管控水平的同时，也在不断追求推陈出新。万达广场和旅游度假酒店的景观规划体系将继续完善，为丰富旅游度假体验和在商业空间中的购物体验，管控方法不断创新，为推动高品质和不同特色的万达广场景观和旅游度假酒店的景观提供强大的推动力。

of sculptural terrain, aesthetically shaped landscape accessories and amusing patterns or facilities, and deliver a vision-active, complex while harmonious landscape environment and experience via contrast in materials and colors (Fig. 4 and Fig. 5).

The innovation of landscape control focuses on freeing our minds to evolve something new boldly. The control personnel in Landscape Office shall, making full sense of systems and mandatory regulations and integrating landscape system and commercial function, constantly pursue landscape feature consistent with commercial function and full of cultural connotation, and consistently seek diversity between different commercial complexes to avoid homogenization. Landscape design suppliers-a significant link of realizing control philosophy, shall be well informed of our philosophy and requirements of control and their corresponding level in supplier library by holding regular exchange conference for suppliers. To encourage internal competition, Landscape Office, through establishing Supplier Integrated Management Information Platform and entire quantitive evaluation system, intends to stimulate innovation effectively and constantly.

With more colorful commercial activities of Wanda, the landscape planning system of Wanda Plaza is also endowed with ever-growing rich connotation, controlling diverse landscape types consisting of shopping center, city hotel, high-end office building and holiday resort (including establishment and maintenance of ecological environment). While constantly enhancing the internal management and control level, the Landscape Office also strives to evolve something new. Landscape planning system of Wanda Plaza and holiday resorts will continue to be improved, attempting to innovate the control approaches for enriching the holiday experience and shopping experience in commercial space, and to provide a powerful drive for creating high-quality and distinctive landscape for Wanda Plaza and holiday resorts.

DESIGN AND CONTROL OF WANDA COMMERCIAL NIGHTSCAPE LIGHTING: A FLOWER OF ART GROWING IN RATIONALITY AND BLOOMING AS NIGHT FALLS

万达商业夜景照明设计与管控——理性中生长 夜幕下绽放的艺术之花

文 / 万达商业规划研究院照明所所长 黄引达

万达商业规划研究院历经十多年的探索实践，具有鲜明万达品牌特色的夜景照明风格逐步形成，收获了一颗颗硕果。在万达夜景照明的发展史中，蕴藏着理性、高效的推进方式，其核心就是专业的设计与管控。

Through more than a decade of exploration and practice, Wanda Commercial Planning Institute has evolved into a nightscape lighting style with distinctive Wanda brand feature, already bearing fruit. In the history of Wanda nightscape lighting, there is a rational and efficient propulsion mode behind, the core of which rests in professional design and control.

一、万达夜景照明的历史沿革

夜景照明对于营造商业氛围，吸引客流，提高品牌知名度及辨识度，有着不可替代的作用。回顾万达夜景照明，历经初始化、规范化、标准化到多元化四个阶段，其过程也反映着国内商业照明的发展史。

1 初始化阶段（2000-2004）

为万达一代店单店时期，商业业态简单，建筑主要为单体建筑。该阶段夜景照明以投光灯加冷阴极管为主，产品单调手法单一；照明控制采用手动方式或简单的时钟控制（图1~图3）。

（图1）长春重庆路万达广场

2 规范化阶段（2005-2010）

2005年万达商业地产开始发力，由二代店转型到成熟的商业综合体模式三代店，万达步入快速布局，批量复制阶段。随着门店数量大幅增加，建设速度加快，夜景照明必然用规范化来保证高速高质的发展。

该阶段建立了设计供方、工程公司、照明灯具品牌库，从设计施工一体化细分为设计、施工独立。投光灯退居后列，LED产品逐渐占据主导，采用智能控制手段，照明手法形式百花齐放，更具商业氛围（图4~图6）。

3 标准化阶段（2010-2013）

万达商业规划研究院在这一阶段设立了照明所，并由

I. HISTORY OF WANDA NIGHTSCAPE LIGHTING

Nightscape lighting plays an irreplaceable role in creating commercial atmosphere, attracting customer flow and enhancing brand awareness & recognition. In retrospect, Wanda nightscape lighting undergoes initialization phase, normalization phase, standardization phase and diversification phase, the process of which mirrors the history of domestic commercial lighting as well.

1 INITIALIZATION PHASE (2000-2004)

Falling into the Wanda Generation I shop period with simple commercial activities and single building, the nightscape lighting then mainly consisted of spotlights

（图2）青岛台东万达广场

with cold-cathode tube, monotonous in both product and technique, and manual or simple clock lighting control was applied (Fig. 1~Fig. 3).

2 NORMALIZATION PHASE (2005-2010)

In 2005, Wanda commercial property was tapped to transform to the mature commercial complex mode-Generation III shop from Generation II shop, and stepped into the quick layout and bulk copy phase. Along with the sharp increase of shops and quickening construction speed, normalization is sure to be applied to nightscape lighting for a high-speed and high-quality development.

In this phase, design supplier, engineering company, lighting fixtures brand library had established;

（图3）南昌八一万达广场

（图4）上海五角场万达广场

（图5）合肥包河万达广场

（图6）宁波鄞州万达广场

（图7）泉州浦西万达广场

（图8）成都金牛万达广场

（图9）温州龙湾万达广场

independent design and construction emerged from the original integrated form; LED products replaced spotlights to gradually hold dominant position; thanks to intelligent control, lighting forms tended to be flourishing, endowing a stronger commercial atmosphere (Fig. 4~Fig. 6).

其牵头全面修订、编制了相关规范及专业技术标准，万达商业夜景照明步入标准化发展阶段，通过标准化设计标准和管控流程，创作出更具商业活力、艺术感染力的产品（图7~图9）。

3 STANDARDIZATION PHASE (2010-2013)

In this phase, Wanda Commercial Planning Institute set up Lighting Office to lead the comprehensive revision and preparation work of relevant specifications and professional skill standards, which declared Wanda commercial nightscape lighting had stepped into the standardization phase. During this period, with standardized design criterion and control process, commodities with stronger commercial vigor and aesthetic appeal were produced (Fig. 7~Fig. 9).

4 多元化阶段（2013–至今）

随着万达商业地产的多元化发展，夜景照明亦呈现出多元化、国际化趋势，建筑媒体照明特点日渐突出，并成为万达商业夜景照明品牌式风格。

4 DIVERSIFICATION PHASE (2013-TO DATE)

Accompanying the diversified development of Wanda commercial property, diversified and international trend is also found in its nightscape lighting. During this phase, architectural media lighting becomes prominent and serves as a brand style peculiar to Wanda commercial nightscape lighting.

2013年出现了“不可复制”的巅峰之作——武汉汉街万达广场，其获得国内外照明界的高度认可，荣膺有“照明奥斯卡奖”之称的2014年国际照明设计师设计最高奖——卓越奖（图10）。其他代表作品见图11和图12。

In 2013, Wuhan Han Street Plaza stood out as a stunning masterpiece that cannot be copied. It gained high recognition in the lighting industry at home and abroad, and awarded Award of Excellence (also the Oscar in Lighting), the top design award of 2014 international lighting designers (Fig. 10). See Fig. 11 and Fig. 12 for other masterpieces.

万达商业规划研究院照明所作为夜景照明专业技术归口管理部门，对夜景照明设计进行归口管理，涵盖了用地内所有建筑室外夜景照明：从商业综合体到城市酒店、度假酒店，写字楼，住宅和底商，室外步行街；从照明方案到施工图，到开业及运营动画设计。实现了设计、建造和运营的全程专业管控。管控内容涉及方方面面，下面仅从万达夜景照明属性展开，对管控的核心进行阐述。

As the centralized management department for nightscape lighting professional skills, the Lighting Office of Commercial Planning Institute, assuming the centralized management work for nightscape lighting design, covers all architectural exterior lighting within the land, ranging from commercial complex to city hotel, resort hotel, office building, residence and floor traders, exterior pedestrian street, and from lighting scheme to construction drawing, animation design of opening and operation. It, thus, implements professional control all the way through design, construction and operation. The control is involved in all aspects, while next, we just start from the attribute of Wanda nightscape lighting to further illustrate the core of control.

二、万达夜景照明的属性

万达夜景照明准确用语应该是万达商业地产夜景照明。商业属性强调对环境氛围的时尚化、娱乐化渲染，对顾客购买欲望的诱导。地产属性则强调成本控制和利润最大化。这一属性理智地抑制了在商业属性

II. ATTRIBUTE OF WANDA NIGHTSCAPE LIGHTING

Wanda nightscape lighting, to be exact, shall be

（图10）武汉汉街万达广场

（图11）长沙开福万达广场

（图12）大连高新万达广场

中对氛围渲染无节制的追求和资金投入。商业属性和地产属性结合，是对夜景照明品质的要求，追求夜景照明对投资回报利润的最大化，和对万达品牌价值的提升，对客流吸引力的诉求。这也与夜景照明自身固有的艺术属性相契合；同时万达属性又附加了对项目的强大的计划控制和对安全的高度重视。

三、万达夜景照明设计管控核心

安全、品质、计划、成本，是万达夜景照明设计管控的四大核心。

1 安全

夜景照明涉及消防、电气、结构、灯具、施工及后期维护安全多个方面，一个典型的万达广场夜景照明有近10万套灯具、10万米配电管线，安全管控责任重大。

为此我们制定了一系列安全标准、制度、管控体系。如《万达消防及安全专篇》、《安全设计管控要点》、《施工图评审会会议标准》、《安全设计制度汇编》等，都是安全管控的有力武器；机电、建筑（幕墙）、照明专业的安全专项施工图审查，评审会会审机制和“安全隐患填报模块”等，一同筑起安全的堤坝。

2 品质

每一座万达广场都承载着区域地标的使命，要求夜景照明设计持续发掘特色主题、提炼当地文化元素并融入夜景视觉表达。同时基于建筑材料、建筑构造的创新应用，照明和建筑紧密结合，对建筑的二次表现，进行多维空间的探索。综合业态的整合、照明产品多元化和控制系统智能化，成为万达夜景照明的技术和效果的创新点。

创新的前提是标准化，《夜景照明设计管控要点》针对施工图设计重点、要点，对签批效果忠实实现进行有效管控；《施工图评审会会议标准》保证各系统、各专业对施工图品质进行综合评审。第三方专家审图制度给品质保证提供了双保险。

Wanda commercial property nightscape lighting. Its commerce attribute emphasizes the fashionable and entertaining embellishment of ambience, and induction of customers' purchasing desire, its property attribute, however, focuses on cost control and maximized profit, which sanely inhibits the immoderate pursuit of ambience embellishment and capital infusion incurred by its commerce attribute. Therefore, the combination of the two attributes is to the quality requirements of nightscape lighting, to the profit maximization of investment, to the enhancement of Wanda brand value and to the appeal of customer attraction. Also, it echoes with the aesthetic attribute rooted in nightscape lighting, while Wanda attribute endows the project an overwhelming planning control and grave concern to safety.

III. CORE OF CONTROL FOR WANDA NIGHTSCAPE LIGHTING DESIGN

1 SAFETY

Nightscape lighting covers fire control, electrical engineering, structure, lighting fixtures, construction and post-maintenance of safety, etc. A typical nightscape lighting of Wanda plaza is furnished with nearly 100, 000 sets of lighting fixtures, 100, 000 m distribution pipeline, thus the responsibility in terms of safety control is extremely heavy.

To this end, we formulate a series of safety standards, systems and control systems, such as *Special Article of Wanda Fire Control and Safety*, *Control Gist of Safety Design*, *Meeting Standards of Construction Drawing Review Meeting* and *Corpus of Safety Design System*, all of which serve as powerful weapons in safety control. Together with safety-specific construction drawing review on electromechanical, architectural (curtain wall) and lighting specialties, joint hearing mechanism of review meeting and "Potential Risk Reporting Module", a safety dam is built thereby.

2 QUALITY

Each Wanda plaza commits the mission of carrying regional landmark, which calls for insistent discovery of special subject in nightscape design, and refinement of the local culture element for incorporating into the visual expression of nightscape. At the same time,

施工质量是品质保证：样板段验收和施工灯具封样，中期检查，规划验收保证了施工图的完美落地。夜景照明效果检查是不同于其他效果专业的另一道保险。

3 计划

计划执行是集团迅速发展的强力保障，模块化系统成为工作的有力工具。照明专业在综合体模块中有四个发起节点，分别是夜景方案确定、夜景给幕墙提资、第一版施工图及灯具样册封样移交、第二版施工图移交。在模块中照明专业还有二个接收节点，分别是幕墙及夜景照明样板段验收和夜景照明施工效果检查。

4 成本

《2013版建造标准》、《集团成本管控办法及操作指引》、《定额设计标准》、《万达广场夜景照明设计选型标准灯具库》等成本标准，配合效果方案、施工图测算、设计变更制度中成本一票否决，具有鲜明的地产特色，对成本控制切实有效。

夜景照明这一璀璨的艺术之花，在各项标准、制度的理性土壤中生长，在全国近百个万达广场上空的夜色中绽放。

based on the innovative application of the building material and building structure, integration of lighting and architecture, and secondary interpretation of the building, exploration of multidimensional space shall be implemented. Integration of comprehensive commercial activities, diversification of lighting products and intelligent control system serve as the innovative performance in terms of technology and effects of Wanda nightscape lighting.

Standardization is a prerequisite for innovation: *Control Gist of Nightscape Lighting Design*, targeting at design focus and gist in construction drawing, exerts effective control on faithful implementation; *Meeting Standards of Construction Drawing Review Meeting* guarantees a comprehensive evaluation of construction drawing quality by each system and specialty; the system of drawing approval by third party experts further ensures the quality.

Construction quality assures quality: sample section acceptance & sealed samples of construction lighting fixtures, intermediate inspection, planning acceptance ensure the perfect achievement of construction drawing. The inspection of nightscape lighting effect is another insurance distinguishing it from other effect specialty.

3 PLAN

Implementation of plan is the strong guarantee for booming of the group, during which modular system acts as a powerful tool. Lighting specialty has four initiating nodes in complex modules, namely, confirmation of nightscape scheme, data provision to curtain wall by nightscape, handover of Version I construction drawing & sealed sample of lighting fixtures catalogue, and handover of Version II construction drawing. Besides, two receiving nodes are included in modules, being curtain wall & nightscape lighting sample section acceptance and inspection of construction effect of nightscape lighting.

4 COST

In terms of cost, it is of distinctive property feature and effective to cost control, as it adheres to *Construction Standard 2013*, *Measures and Operational Guideline of Group Cost Control*, *Quota Design Standard*, and *Standard Lighting Fixture Library of Nightscape Lighting Design and Selection of Wanda Plaza*, and coordinates with effect scheme, reckoning of construction drawing and one-vote veto of cost in design alteration system.

Nightscape lighting-Shining Flower of Art, grows in rational soil nurtured by diverse standards and systems, and blooms in the curtain of darkness on hundreds of domestic Wanda Plazas.

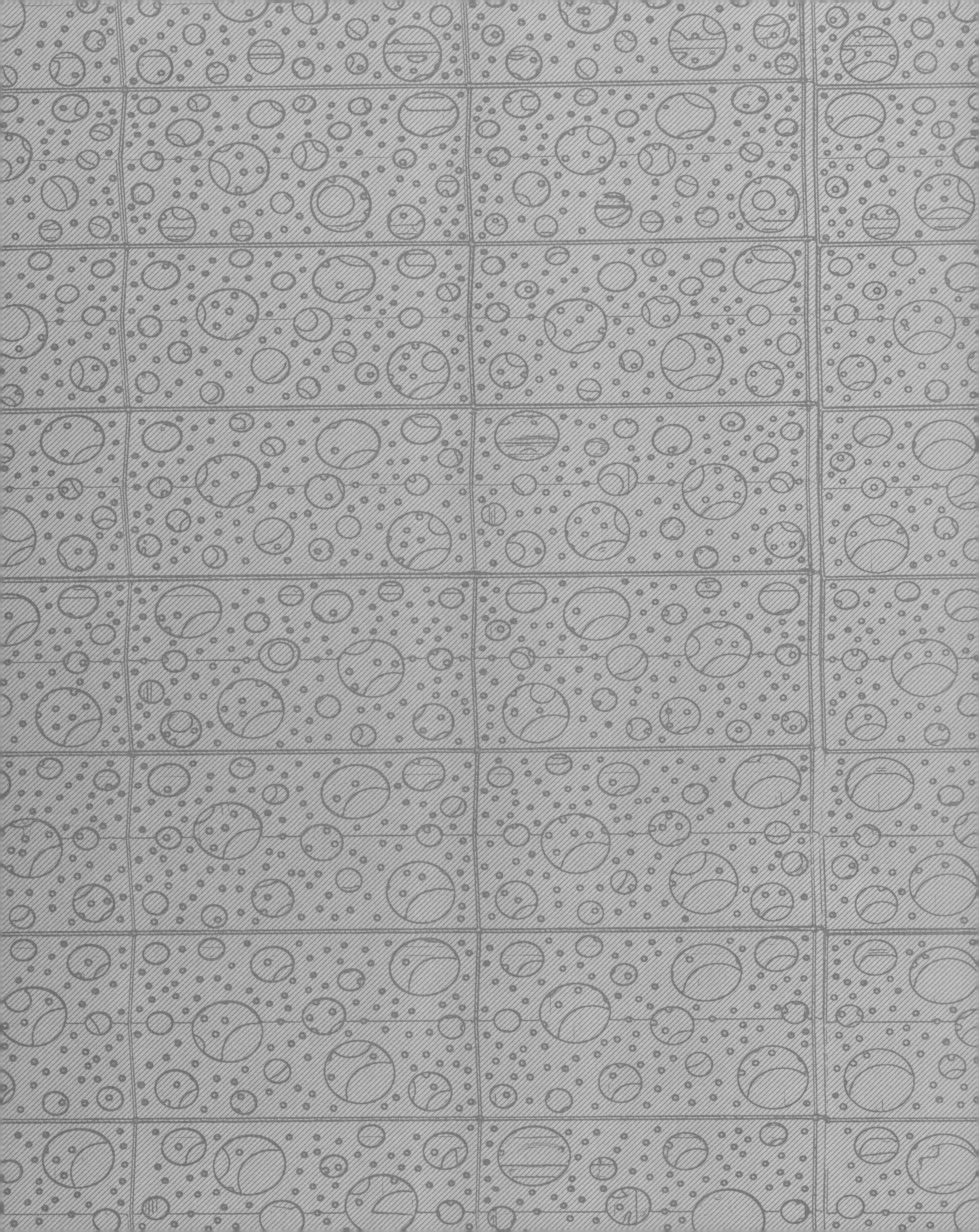

PART

E

WANDA COMMERCIAL PLANNING

INDEX OF PROJECT

项目索引

INDEX OF WANDA PLAZAS
万达广场索引

WUHAN HAN STREET WANDA PLAZA
汉街万达广场
2013 / 09 / 28

外立面设计	UN Studio
幕墙设计	奥雅纳、金星卓宏
内装设计	UN Studio
景观设计	北京易兰建筑规划设计有限公司
夜景照明设计	北京市建筑设计研究院有限公司
导向标识设计	北京博雅空间城市形象设计有限公司
弱电智能化设计	北京国安电气有限责任公司

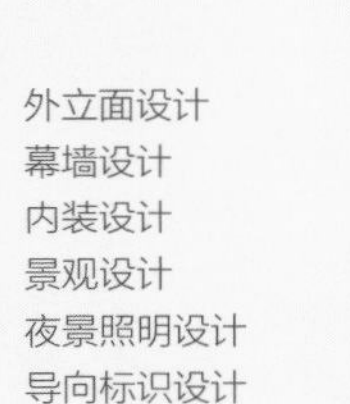

DALIAN HIGH-TECH WANDA PLAZA
大连高新万达广场
2013 / 05 / 25

外立面设计	北京华雍汉维建筑咨询有限公司
幕墙设计	北京市金星卓宏幕墙工程有限公司
内装设计	北京清尚环艺建筑设计院有限公司
景观设计	华东建筑设计研究院有限公司
夜景照明设计	栋梁国际照明设计（北京）中心有限公司
导向标识设计	北京天树文化艺术传播有限责任公司
弱电智能化设计	大连理工科技有限公司

YIXING WANDA PLAZA
宜兴万达广场
2013 / 05 / 31

外立面设计	上海汉米敦建筑设计有限公司
幕墙设计	北京市金星卓宏幕墙工程有限公司
内装设计	北京清尚环艺建筑设计院有限公司
景观设计	中国建筑设计研究院
夜景照明设计	北京三色石环境艺术设计有限公司
导向标识设计	北京艺同博雅企业形象设计有限公司
弱电智能化设计	北京益泰牡丹电子工程有限公司

SHENYANG OLYMPIC WANDA PLAZA
沈阳奥体万达广场
2013 / 07 / 26

外立面设计	赫尔佐格合伙人工作室
幕墙设计	中国建筑科学研究院
内装设计	北京清尚环艺建筑设计院有限公司
景观设计	笛东联合（北京）规划设计顾问有限公司
雕塑及中国印广场景观设计	沈阳莱瑞凯斯装饰工程有限公司
夜景照明设计	北京三色石环境艺术设计有限公司
导向标识设计	北京视域四维设计有限公司
弱电智能化设计	上海中电电子系统工程有限公司

XIAMEN JIMEI WANDA PLAZA
厦门集美万达广场
2013 / 06 / 08

外立面设计	厦门大学建筑设计研究院
幕墙设计	厦门开联装饰工程有限公司
内装设计	深圳市三九装饰工程有限公司
景观设计	福建泛亚远景环境设计工程有限公司
夜景照明设计	深圳市标美照明设计工程有限公司
导向标识设计	北京视域四维城市导向系统规划设计有限公司
弱电智能化设计	北京国安电气工程有限公司

WUXI HUISHAN WANDA PLAZA
无锡惠山万达广场
2013 / 06 / 21

外立面设计	华凯派特建筑设计（北京）有限公司
幕墙设计	北京市金星卓宏幕墙工程有限公司
内装设计	广东省集美设计工程公司 浙江金鼎建筑装饰工程有限公司
景观设计	上海赛特康斯景观设计咨询有限公司
夜景照明设计	北京市建筑设计研究院有限公司
导向标识设计	北京视域四维城市导向系统规划设计有限公司
弱电智能化设计	北京益泰牡丹电子工程有限公司

DONGGUAN CHANG'AN WANDA PLAZA
东莞长安万达广场
2013 / 07 / 20

外立面设计	北京华雍汉维建筑设计有限公司
幕墙设计	厦门开联装饰工程有限公司
内装设计	广东省集美设计工程有限公司
景观设计	北京中建建筑设计有限公司上海分公司
夜景照明设计	北京市建筑设计研究院有限公司
导向标识设计	广育德标识设计有限公司
弱电智能化设计	上海智信世创智能系统集成有限公司

CHANGCHUN KUANCHENG WANDA PLAZA
长春宽城万达广场
2013 / 08 / 16

外立面设计	北京赫斯科建筑设计咨询有限公司
幕墙设计	深圳金粤幕墙装饰工程有限公司(杭州分公司)
内装设计	上海浦东建筑设计研究院有限公司
景观设计	中国建筑设计研究院
夜景照明设计	深圳普莱思照明设计顾问有限责任公司
导向标识设计	深圳上行线设计有限公司
弱电智能化设计	北京国安电气工程有限公司

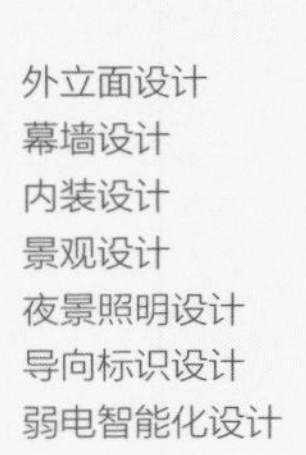

HARBIN HAXI WANDA PLAZA
哈尔滨哈西万达广场
2013 / 09 / 13

外立面设计	上海鼎实建筑设计有限公司
幕墙设计	深圳金粤幕墙装饰工程有限公司
内装设计	北京齐信装饰工程有限公司
景观设计	上海兴田建筑工程设计事务所（普通合伙）
夜景照明设计	北京市建筑设计研究院有限公司
导向标识设计	北京视域思维城市导向系统规划设计有限公司
弱电智能化设计	北京创新迪克系统集成技术有限公司

CHANGSHA KAIFU WANDA PLAZA
长沙开福万达广场
2013 / 09 / 27

外立面设计	德国 KSP尤根-恩格尔建筑师国际有限公司
	北京华雍汉维建筑设计有限公司
幕墙设计	北京市金星卓宏幕墙工程有限公司
内装设计	深圳市三九装饰工程有限公司
	北京联合金玉商业管理有限公司
景观设计	深圳市致道思维筑景设计有限公司
夜景照明设计	北京市建筑设计研究院有限公司
导向标识设计	北京广育德标识制造有限公司
弱电智能化设计	中国通广电子公司

FUSHUN WANDA PLAZA
抚顺万达广场
2013 / 10 / 25

外立面设计	上海新外建工程设计与顾问有限公司
幕墙设计	北京市金星卓宏幕墙工程有限公司
内装设计	深圳市三九装饰工程有限公司
景观设计	华东建筑设计研究院有限公司
夜景照明设计	北京市建筑设计研究院有限公司
导向标识设计	北京视域思维城市导向系统规划设计有限公司
弱电智能化设计	大连理工科技有限公司

NINGBO YUYAO WANDA PLAZA
宁波余姚万达广场
2013 / 11 / 01

外立面设计	北京五合国际建筑设计咨询有限公司
幕墙设计	厦门开联装饰工程有限公司
内装设计	北京清尚环艺建筑设计院有限公司
	杭州友昌装饰设计有限公司
景观设计	华东建筑设计研究院有限公司
夜景照明设计	上海易照景观设计有限公司
导向标识设计	北京艺同博雅企业形象设计有限公司
弱电智能化设计	上海智信世创智能系统集成有限公司

XI'AN DAMING PALACE WANDA PLAZA
西安大明宫万达广场
2013 / 11 / 22

外立面设计	上海联创建筑设计有限公司北京分公司
幕墙设计	深圳蓝波幕墙及光伏工程有限公司
内装设计	广东省集美设计工程公司
	择思设计顾问（深圳）有限公司和
	深圳市中深建装饰设计工程有限公司联合体
景观设计	深圳文科园林股份有限公司
夜景照明设计	深圳市千百辉照明工程有限公司
导向标识设计	北京艺同博雅企业形象设计有限公司
弱电智能化设计	上海中电电子系统工程有限公司

BENGBU WANDA PLAZA
蚌埠万达广场
2013 / 11 / 29

建筑设计	中建国际（深圳）建筑设计顾问有限公司
外立面设计	艾奕康建筑设计（深圳）有限公司
幕墙设计	北京金星卓宏幕墙工程有限公司
内装设计	北京城建长城装饰工程有限公司
景观设计	上海兴田建筑工程设计事务所（普通合伙）
夜景照明设计	北京三色石环境艺术设计有限公司
导向标识设计	北京视域四维城市导向系统规划设计有限公司
弱电智能化设计	上海智信世创智能系统集成有限公司

XUZHOU YUNLONG WANDA PLAZA
徐州云龙万达广场
2013 / 12 / 06

外立面设计	HPP International Planungsgesellschaft mbH
幕墙设计	厦门开联装饰工程有限公司
内装设计	北京清尚环艺建筑设计院有限公司
	江苏建设控股集团有限公司
景观设计	华东建筑设计研究院有限公司
夜景照明设计	深圳市千百辉照明工程有限公司
导向标识设计	北京视域四维城市导向系统规划设计有限公司
弱电智能化设计	上海中电电子系统工程有限公司

DANDONG WANDA PLAZA
丹东万达广场
2013 / 12 / 20

立面设计	艾奕康建筑设计（深圳）有限公司
幕墙设计	北京市金星卓宏幕墙工程有限公司
室内设计	上海帕莱登建筑景观咨询有限公司
导向标识设计	北京视域四维城市导向系统规划设计有限公司
景观设计	铿晓设计咨询（上海）有限公司
夜景照明设计	北京三色石环境艺术设计有限公司
弱电智能化设计	华体集团有限公司

NANJING JIANGNING WANDA PLAZA
南京江宁万达广场
2013 / 12 / 21

外立面设计	北京东方国兴建筑设计有限公司
幕墙设计	上海旭密林幕墙有限公司
内装设计	北京清尚环艺建筑设计院有限公司
	深圳市厚夫设计顾问有限公司
景观设计	上海帕莱登建筑景观咨询有限公司
夜景照明设计	深圳市千百辉照明工程有限公司
导向标识设计	北京艺同博雅企业形象设计有限公司
弱电智能化设计	上海中电电子系统工程有限公司

CHONGQING WANZHOU WANDA PLAZA
重庆万州万达广场
2013 / 07 / 05

外立面设计	上海新外建工程设计与顾问有限公司
幕墙设计	北京市金星卓宏幕墙工程有限公司
内装设计	北京清尚环艺建筑设计院有限公司
景观设计	华东建筑设计研究院有限公司
夜景照明设计	栋梁国际照明设计（北京）中心有限公司
导向标识设计	北京广育德视觉技术股份有限公司
弱电智能化设计	深圳市标盛科技投资有限公司

INDEX OF WANDA HOTELS
万达酒店索引

WANDA REALM WUHAN
武汉万达嘉华酒店
2013 / 09 / 01

外立面设计	贝加艾奇（上海）建筑设计咨询有限公司
幕墙设计	北京市金星卓宏幕墙工程有限公司
景观设计	华东建筑设计研究院有限公司
夜景照明设计	深圳市千百辉照明工程有限公司

WANDA VISTA TIANJIN
天津万达文华酒店
2013 / 09 / 25

外立面设计	华凯建筑设计（上海）有限公司
幕墙设计	深圳蓝玻幕墙及光伏工程有限公司
内装设计	香港郑中设计事务所
景观设计	棕榈园林股份有限公司
夜景照明设计	北京三色石环境艺术设计有限公司司
导向标识设计	视语设计顾问（亚洲）有限公司
弱电智能化设计	北京益泰牡丹电子工程有限公司

WANDA REALM NANCHANG
南昌万达嘉华酒店
2013 / 12 / 13

外立面设计	北京赫斯科建筑设计咨询有限公司
幕墙设计	深圳金粤幕墙装饰工程有限公司（杭州分公司）
内装设计	上海浦东建筑设计研究院有限公司
景观设计	中国建筑设计研究院
夜景照明设计	深圳市千百辉照明工程有限公司
导向标识设计	深圳上行线设计有限公司

WANDA REALM YINCHUAN
银川万达嘉华酒店
2013 / 12 / 18

外立面设计	豪斯泰勒张思图德建筑设计咨询（上海）有限公司
幕墙设计	北京市金星卓宏幕墙工程有限公司
内装设计	铿晓设计咨询（上海）有限公司
景观设计	华汇工程设计集团股份有限公司上海分公司
夜景照明设计	上海译格照明设计有限公司
导向标识设计	北京广育德视觉技术有限公司

PARK HYATT CHANGBAISHAN
长白山柏悦酒店
2013 / 09 / 15

外立面设计	美国 Zehren and Associates, Inc.
幕墙设计	北京市金星卓宏幕墙工程有限公司
内装设计	印度尼西亚 JAYA& ASSOCIATES 香港郑中设计事务所
景观设计	艾奕康建筑设计（深圳）有限公司
夜景照明设计	新加坡 Project Lighting Design 北京市建筑设计研究院有限公司
导向标识设计	Corlette Design 北京视域四维城市导向系统规划设计有限公司

HYATT REGENCY CHANGBAISHAN
长白山凯悦酒店
2013 / 09 / 15

外立面设计	美国 Zehren and Associates, Inc.
幕墙设计	北京市金星卓宏幕墙工程有限公司
内装设计	美国 HBA 设计公司 赫氏 STUDIO
景观设计	艾奕康建筑设计（深圳）有限公司
夜景照明设计	新加坡 Project Lighting Design 北京市建筑设计研究院有限公司
导向标识设计	Corlette Design 北京视域四维城市导向系统规划设计有限公司

WANDA IBIS STYLES HOTELS CHANGBAISHAN / WANDA HOLIDAY INN EXPRESS CHANGBAISHAN
长白山万达宜必思·尚品酒店 / 长白山万达智选假日酒店
2013 / 11 / 08

外立面设计	万达商业规划研究院有限公司酒店所
幕墙设计	北京市金星卓宏幕墙工程有限公司
内装设计	深圳极尚　深圳毕路德　苏州金螳螂
景观设计	宝佳丰（北京）国际建筑景观规划有限公司
夜景照明设计	上海译格照明设计有限公司
导向标识设计	Corlette Design 北京视域四维城市导向系统规划设计有限公司

2013 万达商业规划

持有类物业 下册 VOL.2

WANDA COMMERCIAL PLANNING 2013
PROPERTIES FOR HOLDING

朱其玮 吴绿野 王群华 叶宇峰 兰峻文 张涛 黄勇
赖建燕 孙培宇 刘婷 张琳 苗凯峰 徐小莉 尚海燕
李文娟 安竞 马红 曹春 侯卫华 张振宇 范珑
谷建芳 张振宇 雷磊 王鑫 李彬 张飚 毛晓虎 莫鑫
都晖 刘江 蓝毅 郝宁克 屈娜 冯腾飞 张宝鹏 邵汀潇
万志斌 孙佳宁 袁志浩 阎红伟 吴迪 徐立军 王雪松
张立峰 陈维 谢冕 刘杰 党恩 高振江 沈余 孙海龙
李昕 李海龙 黄引达 孙辉 周澄 齐宗新 刘冰 潘立影
杨艳坤 程欢 邓金坷 康斌 刘易昆 李浩然 李江涛
钟光辉 张宁 张宇 黄春林 黄国辉 耿大治 刘阳
刘佩 石路也 孟祥宾 张洋 章宇峰 陈杰 冯志红
谷强 李小强 葛宁 张鹏翔 田中 虞朋 康宇 王治天
朱岩 董根泉 任腾飞 宫赫谣 王吉 沈文忠 张珈博
刘洋 胡存珊 马逸均 李光 郭晨光 朱迪 王锋 谢杰
李志华 宋锦华 方文奇 刘锋 秦鹏华 杨东 李涛
凌峰 易帆 华锡锋 任洪生 李明泽 刘刚 郭雪峰
陈嘉赟 孔新国 赵洪斌 刘志业 冯董 黄路 曹彦斌
张剑锋 周德 李易 肖敏 段堃 闫颇 唐杰 刘潇
黄川东 熊厚 张雪晖 董明海 李卓东 王静 王昉
谢云 李捷 关发扬 庞博 任意刚 张争 辛欣 傅博
赵陨 杨春龙 顾梦炜 姜云娇 江智亮 白宝伟 王凤华
李健 卫立新 庞庆 何志勇 宋永成 谭訸 卜少乐
高杉楠 韩冰 刘海洋 高峰 王睿麟 王宝柱 野天星
王瑶 葛朗 张佳 王晓昉 曹国峰 李常春 徐春辉
王永磊 于崇 张勰 杨汉国 王文广 张永战 李晓山
罗冲 张旭 高达 赵晓萌 方伟 刘俊 陈海亮 康冠军
晁志鹏 邹洪 郑鑫 周永会 陈志强 陈涛 张宇楠
张绍哲 刘安 全永强 康兴梁 林彦 路清淇 陈晓州
白宇 汤英杰 钱昆 白夜 崔勇 陈理力 刘昕 韦云
杨华 金柱 马辉 杨娜 王朝忠 罗琼 洪斌 刘晓波
赵宁 韩博 徐广揆 张烁君 金博 魏大强 程波 马骁
王鹏 柏久绪 朱广宇 蒲峰 杜晶晶 汤钧 冯科力
主佳 张浩 李扬 刘佳 青云富 王燕 田杰 熊伟 董丽梅
曾明 戎帅 陆峰 李峥 莫力生 李楠 屠波 王巍 吴昊
杨洪海 王惟 刘子瑜 李兵 李树靖 杨宜良 宋樱樱
刘冰 石紫光 牛晋华 程鹏 李为状 王力平 秦彬 王宁
（以入职先后为序）

萬達商業規劃

IMAX